# THE LAND-USE/ TRANSPORT SYSTEM

*Analysis and Synthesis*

# THE LAND-USE/ TRANSPORT SYSTEM

*Analysis and Synthesis*

**W. R. BLUNDEN**

PROFESSOR OF TRAFFIC ENGINEERING
UNIVERSITY OF NEW SOUTH WALES, AUSTRALIA

*Sometime visiting Professor of Transportation Engineering*

UNIVERSITY OF CALIFORNIA, U.S.A.
UNIVERSITY OF BIRMINGHAM, U.K.
UNIVERSITY OF BRADFORD, U.K.
CHULALONGKORN UNIVERSITY, BANGKOK

PERGAMON PRESS

OXFORD · NEW YORK · TORONTO
SYDNEY · BRAUNSCHWEIG

Pergamon Press Ltd., Headington Hill Hall, Oxford
Pergamon Press Inc., Maxwell House, Fairview Park, Elmsford, New York 10523
Pergamon of Canada Ltd., 207 Queen's Quay West, Toronto 1
Pergamon Press (Aust.) Pty. Ltd., 19a Boundary Street, Rushcutters Bay, N.S.W. 2011, Australia
Vieweg & Sohn GmbH, Burgplatz 1, Braunschweig



First edition 1971
Library of Congress Catalog Card No. 77-171839



*Printed in Great Britain by Bell and Bain Ltd., Glasgow*

**08 016777 2**

## Contents

## 3. Transport Impedance and Traffic Flow



## 4. Mathematical Programming Methods



## 5. Transport System Characteristics and Performance

## 9. Traffic Assignment



## 10. Evaluation Review and Implementation

# Preface

When the School of Traffic Engineering in the University of New South Wales was established in 1956 at Professorial level it was the first such school in the British Commonwealth. It owed its origin to the vision and generosity of the Australian Automobile Association which underwrote its initial operations. Although its activities in the early years were oriented towards the problems of road traffic it soon became obvious that the so-called "traffic problem" could not be defined phenomenogically in any less a context than that resulting from the interaction between land use and transport. Over the years the teaching and research activities of the staff and graduate students has been directed towards the evolution (in fact mutation) of traffic engineering into a discipline for the study of land-use/transport interaction phenomena—hence the title of this work.

This book is based on the course of lectures given by the author in Land Use and Transport Planning. Whilst it would not be right to claim for it any originality in respect of the concept that traffic is the joint consequence of land-use activity and transport capability it is hoped that it breaks new ground in formalizing this intuitively obvious concept. In doing this the author has drawn heavily on the methods and practice of operations research. Although the true power of the quantitative approach is based on the fundamental doctrine of physics—to define, measure and develop conceptual relationships, it is also true that in the broad context of socio-economic phenomena the methods and practice of operations research are an indispensable aid. Originally it was intended to include the sections on queueing theory and mathematical programming as appendices. However, as the work progressed it became apparent that the jargon and the technique of operations research are so interwoven with the formal development of the basic theory of land use and transport interaction that they have been included as full chapters early in the book. Further, the student of systems operations and planning must today achieve some degree of competence with these new mathematical tools and the introductory account given in Chapters 3 and 4 will, it is hoped, stimulate his interest to do so. The practitioner may, however, omit these chapters in a first study of the book.

Although the book is concerned mainly with the methodology of land-use/transport-system analysis a considerable amount of land-use and transport data in the form of tables and graphs has been included. It will not only provide the student with numerical values for many essential constants and parameters for use in assignments and exercises, but should be of value to the practitioner in making preliminary analyses of real problems. The majority of the data supplied have been obtained by graduate students working under the direct supervision of the author. In other cases the source of the data is indicated by appropriate referencing. Some is typical only.

In offering this book to the student of traffic and to others with professional interest in town planning, transport engineering, economics and management, the author gratefully acknowledges the contributions that have been made to his development of the subject by the staff and graduate students of the School of Traffic Engineering and to his many other colleagues at home and abroad. In particular he would record his appreciation of his happy and valuable associations with staff and students at the Universities of California, Birmingham, Bradford and the Asian Institute of Technology in Bangkok.

Finally he would record also his debt to Ellen Doherty and Merellyn Peter for their painstaking work in the preparation of drafts and the final text.

*University of New South Wales, Sydney* W. R. BLUNDEN

CHAPTER 1

# Introduction—Definitive Concepts and Ideas

## THE NATURE OF TRAFFIC

Traffic is a medium of activity. It is all too often considered to be synonymous with transport facilities or communication apparatus and this has led to a variety of one-sided criteria for assessing the needs for, and performance of, transport systems. This notion often sets the transport specialist in opposition to the town planner, the sociologist and economist and all too often leads to the futile "chicken-and-egg" argument as to which comes first—transport or land use.

The truth is that traffic is the joint consequence of land-use potential and transport capability. It is intuitively obvious that neither land use nor transport on its own can cause or generate traffic and traffic is, in fact, the medium in which both find expression. Land use and transport form a closed-loop system. One can, of course, open the loop and study the effect on traffic of imposing changes on one or the other and this is an essential and valuable exercise in the study of traffic. When the loop is closed, however, the system, like any other physical or economic system, will in the long term settle down into a state of equilibrium even though the equilibrium state may be far from optimum.

Many everyday experiences amply validate this basic proposition that both land use and transport are necessary to produce traffic. There would be no traffic to justify the construction of a railway or motorway into the centre of a desert—unless of course a deposit of copper, say, were located there. Similarly there would be little enthusiasm in the boardroom for a proposal to promote a centre for entertainment and recreation in a remote place with no transport links with the outside world.

In practice it would be difficult to imagine extreme situations in which either land-use activity or transport capability were totally absent. Even so, one can appreciate that there would be little point in providing access to a large regional shopping centre via a 20-foot lane or building an elaborate freeway to serve a rubbish dump.

But important now is the point that if both the land use and transport do exist the traffic that comes into existence is determined by the land-use potential and the transport capability. One of the important objectives of planning a land-use/transport system is to ensure that there is an efficient balance between land-use potential and transport capability.

It should be pointed out also that the socio-economic environment in which traffic arises determines the scale of the land use and transport activity. Apart from the imbalance that may occur from mismatching the land use and transport components of a system, there is the problem that both may be mismatched to the environments in which they occur. For example, it would be foolish to contemplate the development of a large supermarket with parking space for 2000 vehicles complete with roads to serve it in a village of 500 inhabitants.

One could not make any claim to originality in respect of any general thesis concerning the mutual interdependence between land use and transport, for at all levels of discussion such a notion finds ready enough acceptance. However, little has been done so far to formalize the concepts involved in order that they may be advanced as some kind of a basic discipline for the many different groups that have professional and lay responsibility for the various aspects of land use and transport plans, operations and policy.

In setting the stage for what follows in this work, it is important to stress that the study of traffic, in the sense that it is a measure of socio-economic activity, must embrace movement of goods as well as persons.

## REPRESENTATION OF A LAND-USE/TRANSPORT SYSTEM

A land-use/transport system may be represented by a spatial array of land-use zones overlaid with a network representing the transport system. Such a system is shown diagrammatically in Fig. 1.1. There are many points of interest connected with this diagram.

The land-use zones should ideally define an area of homogeneous land-use activity, i.e. residential, industrial, commercial, recreational, retailing, and so on. The extent to which such an ideal definition can be realized depends on the "scale" of the problem under consideration. If the ultimate in detail is required, land-use zones must in general be specified on a block-by-block basis. There are many important problems

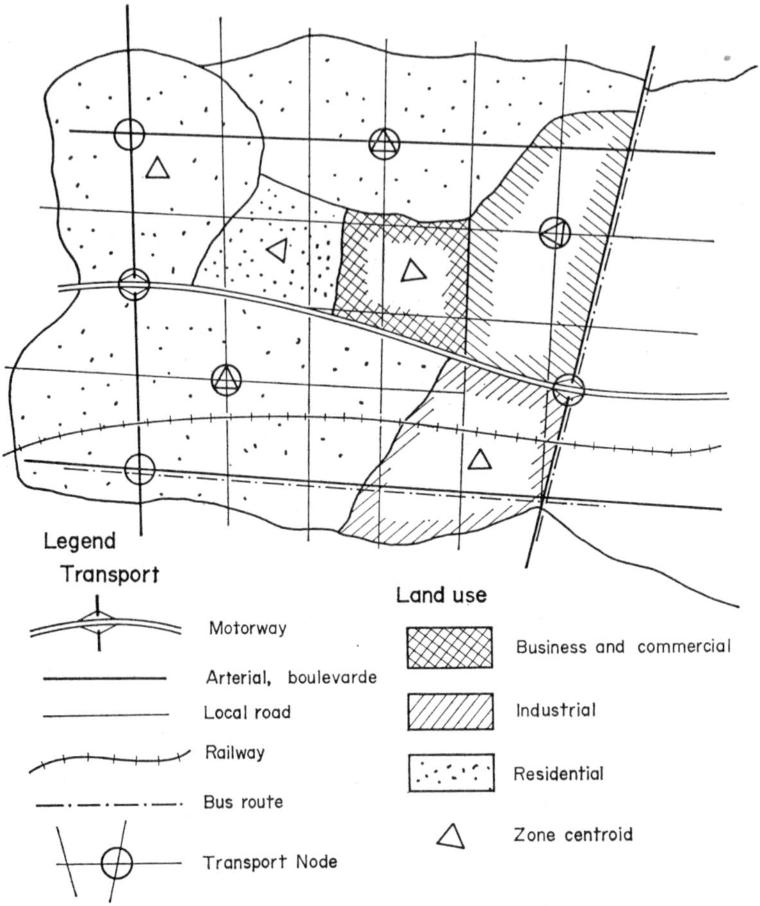


FIG. 1.1. Representation of a land-use/transport system.

of land-use and transport interaction that call for such detail; for example, problems relating to development applications for supermarkets, motels, internal traffic movement in a business centre or industrial estate. On the other hand, the development of an overall land-use/transport plan for a metropolitan area requires extensive aggregation of individual land uses. It is fortunate that at the macroscopic level extended zones do exhibit a great deal of homogeneity and permit the land-use plan of most cities to be divided into some twenty to fifty major zones which permit business centres, industrial estates, residential areas to be characteristically defined.

However, irrespective of the absolute size of a land-use zone, if a quantitative analysis is to be undertaken it is convenient to consider its activity to be concentrated at a point usually known as the zone centroid. In certain situations it is appropriate to regard the zone boundary as a somewhat hypothetical geographical contour which encloses a number of types of land-use activity which are grouped and represented by separate centroids within the zone.

A system of land-use zones may be specified numerically in terms of

(a) the coordinates of its centroid or centroids or the inter-centroid distances;

(b) a measure of the scale of the activity associated with each centroid, e.g. geographical area, number of dwelling units, square feet of retail trading space, office accommodation, number of employees, or the trip ends that emanate from or finish in the zone.

Even before the land-use plan is overlaid with a transport network the potential of the land-use zones and their spatial separation set up the basic transport task. This may be specified as shown in a desire-line matrix as shown in Fig. 1.2. The activity situation represented relates to work trip desires, $T_{od}$, between three residential zones and two work centres. The land-use potentials in this example are resident workers, $R_o$, and the number of jobs, $E_d$, respectively, and the transport capability is inferred from the interzonal travel times, $t_{od}$. This matrix defines the classical Transportation Problem of Linear Programming.

There is no difficulty in specifying the transport network either geometrically or numerically, for in diagrammatic form it is simply a two-dimensional array of links and nodes. Much will be said later about the application of network theory to the study of the flow of traffic in transport networks; our concern here is with the abstraction of real transport networks for analytical purposes. To do this effectively it is necessary to distinguish a hierarchy of transport systems, viz.

(a) the trunk transport network,
(b) the distributor grid,
(c) a reticulation or local access system.

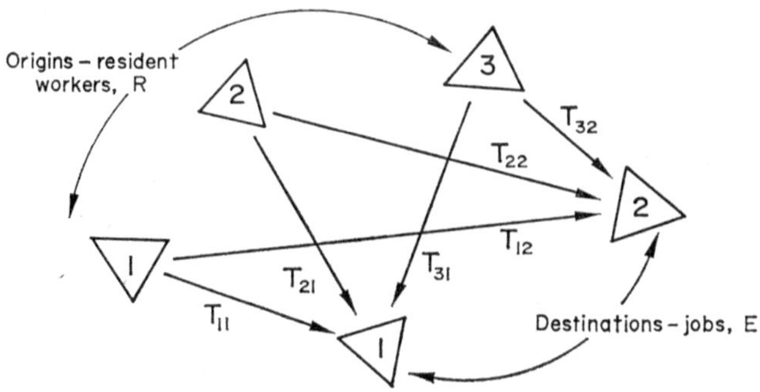


(a) An origin and destination pattern resulting from 3 residential and 2 employment land use zones

| o \ d | 1 | 2 | R's |
|---|---|---|---|
| 1 | $T_{11}$ / $t_{11}, d_{11}$ | $T_{12}$ / $t_{12}, d_{12}$ | $R_1$ |
| 2 | $T_{21}$ / $t_{21}, d_{21}$ | $T_{22}$ / $t_{22}, d_{22}$ | $R_2$ |
| 3 | $T_{31}$ / $t_{31}, d_{31}$ | $T_{32}$ / $t_{32}, d_{32}$ | $R_3$ |
| E's | $E_1$ | $E_2$ | |

(b) Traffic desire line matrix

FIG. 1.2. Matrix representation of a system of land-use zones.

Further, one must specify a range of transport modes—road, rail, sea and air, and most importantly walking or manhandling, for in the ultimate every trip is either a door-to-door or a floor-to-floor event. After all, the climax to man's most remarkable transport feat was to walk on the surface of the moon! It should be noted too that a transport mode may be operated to provide individual or mass movement. Figure 1.1 serves to illustrate these definitive notions of a transport system hierarchy. The motorway and the railway would comprise elements of the trunk transport system of an extended region. The arterials would provide the principal distributor facilities and the other roads would provide access to individual land uses. It should be noted too that the road transport mode provides for bus services on two of the arterial roads.

Although there is a certain amount of interdependence between the functional hierarchy of the transport system and the different modes, it is by no means hard and fast. The trunk transport system may comprise any one or a combination of the first four modes, but not walking. On the other hand, the reticulation network may utilize the walking and road modes, but hardly sea—except of course in Venice.

There will be a need later on to consider in some detail the interrelations implicit in this double hierarchy, particularly because such interrelationships throw into correct perspective the interfaces or transfer facilities between the functional and modal hierarchies—interchanges, railway and bus terminals, airports, docks and, remembering the walking mode, parking.

For analytical purposes it is very convenient to specify the transport network in matrix form. This may be done by constructing a two-dimensional inter-nodal link matrix or a node–link matrix, as illustrated in Fig. 1.3. In the former the elements 1 or 0 indicate the existence or otherwise of a directional link between any pair of nodes $i$ and $j$ and in the latter whether or not a link emanates from a given node. The elements of an array of this type may be used also to specify the length, travel time, cost or capacity of the links of the network.

Before concluding this section on representation of a Land-use/Transport complex, it is of some significance to emphasize again that a transport task is implicit in disposition of land-use zones irrespective of whether or not it is overlaid with a transport network simply because of the spatial separation of the zones themselves. This notion is important because it provides a basis for a later discussion on concept planning.

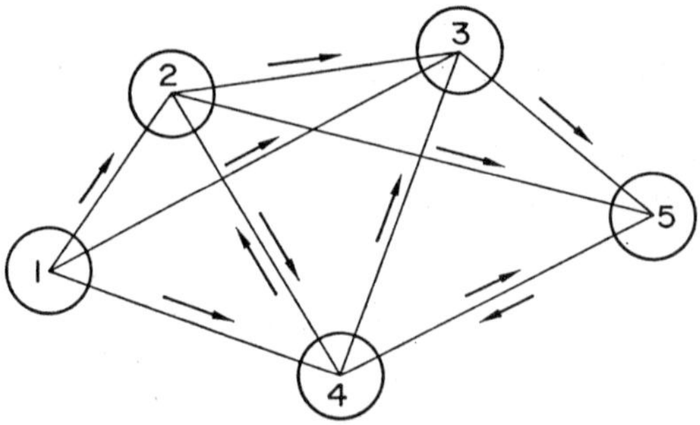


(a) A 5-Node transport network

| i \ j | 1 | 2 | 3 | 4 | 5 |
|---|---|---|---|---|---|
| 1 | – | 1 | 1 | 1 | |
| 2 | | – | 1 | 1 | 1 |
| 3 | | | – | | 1 |
| 4 | | 1 | 1 | – | 1 |
| 5 | | | | 1 | – |

(b) The node-node matrix

| i/j \ ij | 12 | 13 | 14 | 23 | 24 | 25 | 35 | 42 | 43 | 45 | 54 |
|---|---|---|---|---|---|---|---|---|---|---|---|
| 1 | 1 | 1 | 1 | | | | | | | | |
| 2 | -1 | | | 1 | 1 | 1 | | -1 | | | |
| 3 | | -1 | | -1 | | | 1 | | -1 | | |
| 4 | | | -1 | | -1 | | | 1 | 1 | 1 | -1 |
| 5 | | | | | | -1 | -1 | | | -1 | 1 |

(c) The node-link matrix

FIG. 1.3. Representation of a transport network in matrix form.

## THE LAND-USE/TRANSPORT INCIDENCE MATRIX

It is possible to describe the traffic status of the complete Land-use/Transport System by means of a three-dimensional array $A(l, m, n)$ which defines the traffic flow $q(m, n)$ on link $l$ of the transport network. $q(m, n)$ is that part of the traffic-desire line, $n$, between an $o$–$d$ pair that uses the $m$th of the various practical alternative paths or routes between them.

Figure 1.4 illustrates the construction of the matrix for the simple land-use/transport network shown in the accompanying sketch.

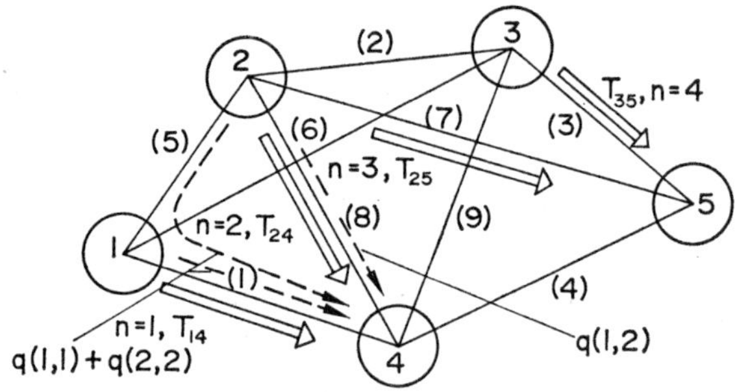


(a) The network

| Travel times | | | $t_1$ | $t_2$ | $t_3$ | $t_4$ | $t_5$ | $t_6$ | $t_7$ | $t_8$ | $t_9$ | $\Sigma t_v$ |
|---|---|---|---|---|---|---|---|---|---|---|---|---|
| q(m,n) | n | $l$ / m | 1 | 2 | 3 | 4 | 5 | 6 | 7 | 8 | 9 | |
| q(1,1) | 1 | 1 | 1 | | | | | | | | | $t_1$ |
| q(1,2) | 2 | 1 | | | | | | | | 1 | | $t_8$ |
| q(2,2) | | 2 | 1 | | | | 1 | | | | | $t_1 + t_5$ |
| q(1,3) | 3 | 1 | | | | | | | 1 | | | $t_7$ |
| q(2,3) | | 2 | | | | 1 | | | | 1 | | $t_4 + t_8$ |
| q(3,3) | | 3 | | 1 | 1 | | | | | | | $t_2 + t_3$ |
| q(1,4) | 4 | 1 | | | 1 | | | | | | | $t_3$ |

(b) Matrix

FIG. 1.4. The land-use/transport incidence matrix.

It is not appropriate at this stage to set up the equations of the system. But it is appropriate to define

(a) the flows on every link of the network that arise from the imposition of the traffic demand represented by the desire lines $T_n$; they may be written down by inspection from simply summing vertically the flow components times the 0, 1 element in the link column, i.e.

$$\sum_m \sum_n A(l, m, n) \,.\, q(m, n) \qquad l = 1, 2, \ldots, L$$

giving

| | | | |
|---|---|---|---|
| on link 1 | $q(1, 1)+q(2, 2)$ | on link 6 | 0 |
| on link 2 | $q(3, 3)$ | on link 7 | $q(1, 3)$ |
| on link 3 | $q(3, 3)+q(1, 4)$ | on link 8 | $q(1, 2)+q(2, 3)$ |
| on link 4 | $q(2, 3)$ | on link 9 | 0; |
| on link 5 | $q(2, 2)$ | | |

(b) the demands can be defined in terms of the flows on the links by summing the $q(m, n)$ column in sections for each value of $n$, i.e.

$$\sum_m q(m, n) \qquad n = 1, 2, \ldots, N$$

giving for desire line $1(T_{14})$ $q(1, 1)$,
" " " " $2(T_{24})$ $q(1, 2)+q(2, 2)$,
" " " " $3(T_{25})$ $q(1, 3)+q(2, 3)+q(3, 3)$,
" " " " $4(T_{35})$ $q(1, 4)$,

(c) the total amount of travel in the system is obtained by summing horizontally the product of matrix elements and length (or travel times) of the links and then adding the row totals vertically, i.e.

$$V = \sum_l \sum_m \sum_n A(l, m, n) t_1 q(m, n)$$

giving $V = t_1 q(1, 1) + t_8 q(1, 2) + (t_1 + t_5) q(2, 2) + t_7 q(1, 3)$
$+ (t_4 + t_8) q(2, 3) + (t_2 + t_3) q(3, 3) + t_3 q(1, 4).$

A HYPOTHETICAL LAND-USE/TRANSPORT SYSTEM—ANZAC CITY

We shall describe here a small but typical land-use/transport system of an urban kind. It is a town of about 67,000 population which we shall call ANZAC City by way of tribute to the valuable contributions to this work made by the many Australian and New Zealand students who have studied together in the School of Traffic Engineering in the University of New South Wales. This hypothetical town on which many student exercises have been carried out originated as Lartsville.[1] It was modified to become U.K. Newtown in the author's Birmingham

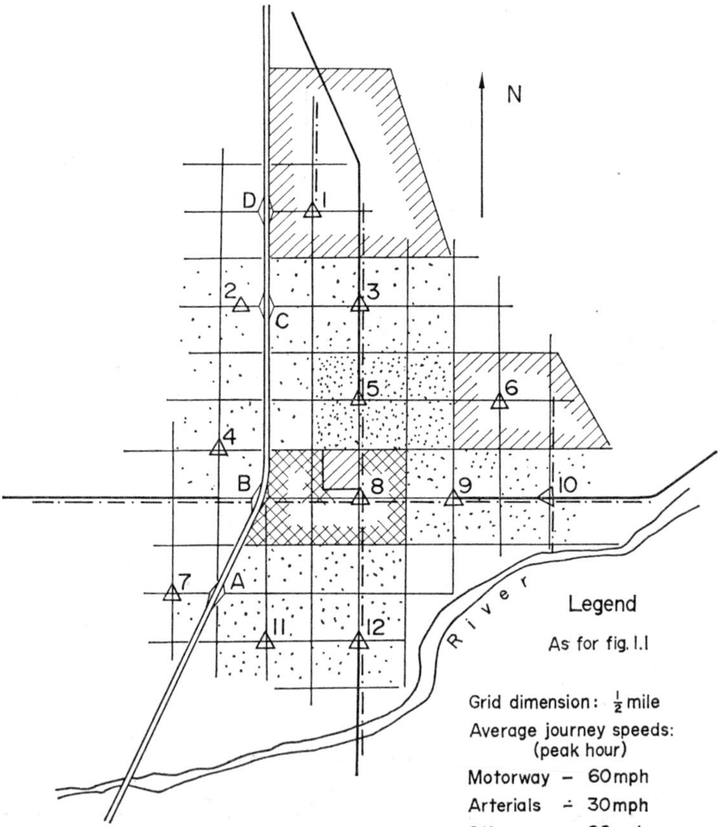


FIG. 1.5. ANZAC City—basic land-use and transport plan.

lectures.[2] ANZAC City has been the guinea pig for a number of new exercises since then but its land-use/transport plan is identical to that of U.K. Newtown—it has been changed in name only in this work!

The land use and transport geometry of ANZAC City is shown in Fig. 1.5. The three work zones 1, 6 and 8 are segregated from the nine residential zones 2, 3, 4, 5, 7, 9, 10, 11, 12. This does not result in loss of generality in analytical evaluations. The major shopping facilities are assumed to be located in zone 8—the Central Business District. It is also the locale for business and commercial activity but there remains in it a small number of industrial-type jobs. The potential for general social intercourse may be identified with the residential zones. Schooling and neighbourhood shopping may be assumed to result in internal traffic only. Data relating to the land-use potential of these zones for inter-zonal traffic activity will be given in the next section.

The transport system for individual and mass transport is road based. The road system currently available has a simple grid pattern built around a north–south state highway and an east–west main road. The various illustrative land-use and transport system analyses to be carried out on ANZAC City will assume that the future road system will incorporate an access-controlled road, generally parallel to the state highway with connections to the surface street system at interchanges A, B, C and D. The bus routes are specified in the legend and it may be assumed that buses will not use the freeway.

If one notes only that the grid is square with a dimension of 0·5 miles, the plan itself gives the essential geometrical route data on the city's transport system. Certain operational characteristics of the transport system need to be specified if one is to be able to determine the travel times between zones. However, the data will be introduced when we discuss transport impedance later in this chapter.

## LAND-USE POTENTIAL AND TRAFFIC GENERATION

Land-use potential is a measure of the scale of socio-economic activity that takes place on a given area of land. Such activity conjures up in one's mind a kaleidoscopic vision of human endeavour—farmers tilling the soil, mine shafts vomiting ore, families in their homes, shoppers burdened with parcels, executives around the board-room table, children at their school desks, workers tending machines, crowds watching football, couples eating in restaurants, planes standing by on the tarmac, cars spread around parking lots, and so on almost indefinitely.

In the phenomenological sense the study of land-use activity provides a common meeting place for a wide range of disciplines—sociology, architecture, demography, economics, physics, mathematics, planning and many more. As yet, however, little progress has been made in formally describing and measuring this complex activity situation. However, a unique property of land use is its ability or potential to "generate" traffic. Workers and raw materials flow onto an industrial estate at certain times and workers and finished products emanate from it at others. Similarly, a residential area attracts and generates workers, shoppers, school children and goods and rubbish. Farms and mines generate grain and ore. In the case of person trips there is a conservation of flow, i.e. as many persons return as go out. In the case of materials there is in general a net flow from producer to consumer—at least in the physical transport sense. Timber moves progressively from the forest via the sawmill to the home site. In like manner, grain and ore make a one-way journey. However, if one could take account in some appropriate way of the relationship between matter and energy there would in fact be some kind of balance achieved between the inflow and outflow. It will be important later to grasp the idea that a zone of land-use activity can be variously a source, a sink or a node with storage capacity. At this point, however, we wish only to establish the idea that once having defined a land-use zone by enclosing land with a geometric boundary, a flux of traffic is established across the boundary contour. In such circumstances it is appropriate to relate the traffic potential of a land use, $L$, and the amount of traffic that will cross its boundary, $G$, in the following manner:

$$G = K \,.\, L.$$

The proportionality "constant" $K$ is in general a function of socio-economic variables. For example, the number of outbound trips per household per day depends on the average family size, the income level, motor-vehicle ownership, social and educational background and the density (not the number) of the houses themselves. The total number of houses in the zone is the appropriate measure of $L$—the traffic potential.

$K$ is in fact the traffic generation parameter or function and converts the arbitrary measure of scale of land-use activity into a real traffic measure.

The average density of the traffic flux across the boundary contour may be written

$$d = KL/S$$

where $S$ is the perimeter of the zone boundary. In practical problems it is instructive to measure the density of the traffic flux in terms of the number of roadway lanes or transport channels that cross the boundary.

TABLE 1.1. MEASURES OF LAND-USE POTENTIAL

| Type of land use | Measure |
|---|---|
| Towns and cities | Population, vehicles |
| Employment zones | |
| Factories<br>Offices | Area, number of employees |
| Shopping centres | Retail trading area and/or parking spaces, number of retail channels |
| Residential | Population, dwelling units, resident workers |
| Recreational | |
| Theatres | Seating capacity |
| Restaurants | Seating capacity |
| Camping sites, beaches, parks | Area, parking spaces |
| Hotels | Number of rooms, capacity of lounges, bars, dining rooms, conference halls |
| Production facilities | |
| Industrial | Output capacity—tons, production units |
| Rural | Acres, carrying capacity—head of stock, bushells |
| Mining | Reserves of ore, fuel |
| Transport terminals | Platforms, berths, parking spaces, area |

One should note, however, that traffic generation is a dynamic phenomenon and it is necessary to define the intensity of traffic generation, $g$, as a function of time and space coordinates. More will be said on this important subject in the chapter dealing with traffic generation. Our object here has been merely to define and establish the notion of traffic generation as a unique measure of a land-use activity. Table 1.1 gives a summary of common measures of land-use potential.

We will conclude this introductory explanation of land-use potential by setting out certain land-use data of relevance to the analyses that will be carried out on ANZAC City. They are given in Table 1.2. The populations of the residential zones and the number of jobs available in the employment zones are appropriate measures. The value of the land is included in this table to provide a basis for certain land-use calculations later on.

TABLE 1.2. LAND-USE DATA FOR ANZAC CITY

| Zone | Population | | Employment | | | | Land value $/acre (000's) |
|---|---|---|---|---|---|---|---|
| | | | Comm. and Retail | | Industrial | | |
| | Pres. | Max. | Pres. | Max. | Pres. | Max. | |
| 1 | | | | | 4500 | 20,000 | 4·0 |
| 2 | 3000 | 12,000 | | | | | 1·0 |
| 3 | 12,100 | 22,000 | | | | | 3·0 |
| 4 | 3800 | 25,000 | | | | | 1·5 |
| 5 | 17,600 | 20,000 | | | | | 4·0 |
| 6 | | | | | 9000 | 15,000 | 6·0 |
| 7 | 1900 | 25,000 | | | | | 1·0 |
| 8 | | | 8000 | 25,000 | 1000 | 2000 | 10·0 |
| 9 | 8800 | 10,000 | | | | | 6·0 |
| 10 | 12,800 | 18,000 | | | | | 5·0 |
| 11 | 2000 | 20,000 | | | | | 5·0 |
| 12 | 5000 | 20,000 | | | | | 5·0 |
| Totals | 67,000 | 172,000 | 8000 | 25,000 | 14,500 | 37,000 | |

TRANSPORT IMPEDANCE AND TRAFFIC FLOW

The impedance of a transport element or system of elements which forms a network is conveniently measured by a suitable function which is synonymous with the "difficulty" of travelling over the link or progressing through the network. It is at first a little surprising that the time or cost of making a journey provides a simple and sufficient measure of transport impedance.

In the strict economic sense the cost measure is not so difficult to accept. However, there is a wealth of experimental evidence to show that in the workaday world time is money, and furthermore in the transport context the cost of a journey is in general proportional to the time it takes.

The impedance of a transport element is not, however, independent of the flow along it. This is a very fortunate circumstance for the nature

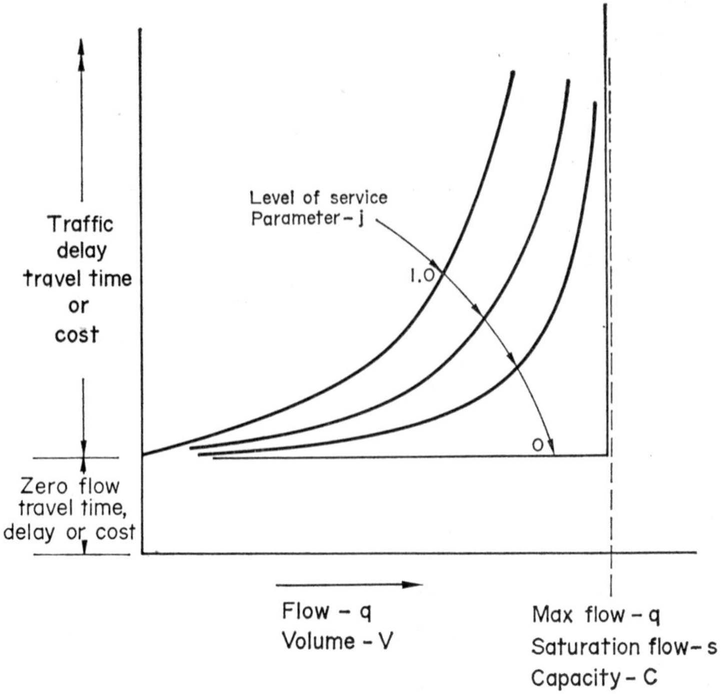


FIG. 1.6. The transport impedance function.

of the relationship between travel time or cost and the traffic flow has a non-linear characteristic which will in general ensure the stability of the land-use transport system.

Figure 1.6 shows the general form of a travel time/flow characteristic. At low flows the impedance of the traffic element has its lowest value. As the flow rises the impedance increases, slowly at first and as the maximum flow level is approached the impedance rises rapidly and in the "steady state" situation is asymptotic to the saturation flow ordinate.

The diagram also indicates that traffic elements with similar zero flow/travel time and maximum flow capability may have differently shaped characteristics. In progressing from left to right in the set illustrated, the individual curves show that the impedance for a given level of flow is reduced; in other words they represent a more efficient flow situation or a higher "level of service". A great deal of discussion will be devoted later on to the travel time/flow properties of traffic elements. At this stage it is pertinent to remark that the travel time/flow characteristic has a place in traffic theory similar to that of Ohms' law in electricity, or Hooke's law in structures, and the basic theory of traffic flow is well founded in the Theory of Queues and Waiting Lines. It is also relevant to note here that this set of characteristic curves has a great universality of application to transport devices of all types—sections of road, intersections, airport runways, loading and unloading docks, ship berths, ticket windows, toll facilities, baggage check-in counters, retail service channels, and so on.

From the above discussion it will be seen that transport impedance depends on

(a) the zero flow travel time, i.e. the travel impedance between two cities will be reduced as the "all stations" gives way to the express and the express in turn to the jet airliner;

(b) the saturation flow rate of a succession of vehicles through or over the transport link;

(c) the level of service factor.

There is a great deal of practical technology associated with the exploitation of this basic impedance concept, particularly if we note that every journey is either a door-to-door or a floor-to-floor event. The travel time for a bus journey from a suburban residence to the city office involves the walk to the bus stop, the waiting time, the routing of the bus service, the number and length of stops, the traffic situation, the walk to the office.

In the more general context of transport operations, i.e. journeys embracing more than one mode and freight as well as passenger movement, the cost of the trip or journey represents a more appropriate and

TABLE 1.3. INTERZONAL TRAVEL TIMES FOR ANZAC CITY, WITH AND WITHOUT PROPOSED MOTORWAY

With motorway (min) / Without motorway (min)

| Zone | 1 | 2 | 3 | 4 | 5 | 6 | 7 | 8 | 9 | 10 | 11 | 12 |
|---|---|---|---|---|---|---|---|---|---|---|---|---|
| 1 | | 4·3/6·3 | 4·5/4·5 | 8·5/11·5 | 6·5/6·5 | 14·0/14·0 | 8·1/17·0 | 13·5/14·5 | 11·5/12·5 | 13·5/14·5 | 9·6/16·0 | 12·6/14·0 |
| 2 | | | 3·75/3·75 | 5·3/5·3 | 5·8/5·8 | 13·3/13·3 | 5·3/10·7 | 10·8/13·8 | 8·8/11·8 | 10·8/13·8 | 6·9/11·3 | 9·9/13·3 |
| 3 | | | | 8·0/8·0 | 2·0/2·0 | 9·5/9·5 | 7·6/13·5 | 10·0/10·0 | 8·0/8·0 | 10·0/10·0 | 9·1/12·5 | 9·5/9·5 |
| 4 | | | | | 6·0/6·0 | 13·5/13·5 | 5·5/5·5 | 11·0/11·0 | 9·0/9·0 | 11·0/11·0 | 7·1/7·5 | 10·1/10·5 |
| 5 | | | | | | 7·5/7·5 | 8·6/11·5 | 8·0/8·0 | 6·0/6·0 | 8·0/8·0 | 10·5/10·5 | 7·5/7·5 |
| 6 | | | | | | | 15·6/18·5 | 15·0/15·0 | 7·0/7·0 | 7·0/7·0 | 17·1/17·5 | 14·5/14·5 |
| 7 | | | | | | | | 10·6/13·5 | 8·6/11·5 | 10·6/13·5 | 4·5/4·5 | 7·5/7·5 |
| 8 | | | | | | | | | 8·0/8·0 | 10·0/10·0 | 12·1/12·5 | 9·5/9·5 |
| 9 | | | | | | | | | | 2·0/2·0 | 10·1/10·5 | 7·5/7·5 |
| 10 | | | | | | | | | | | 12·1/12·5 | 9·5/9·5 |
| 11 | | | | | | | | | | | | 3·0/3·0 |
| 12 | | | | | | | | | | | | |

perhaps more sensitive measure of transport impedance. Due attention will be paid to this matter in Chapter 5.

At this stage it is perhaps sufficient to say that the "cost" of a trip is, to a good measure of approximation, proportional to the total time it is

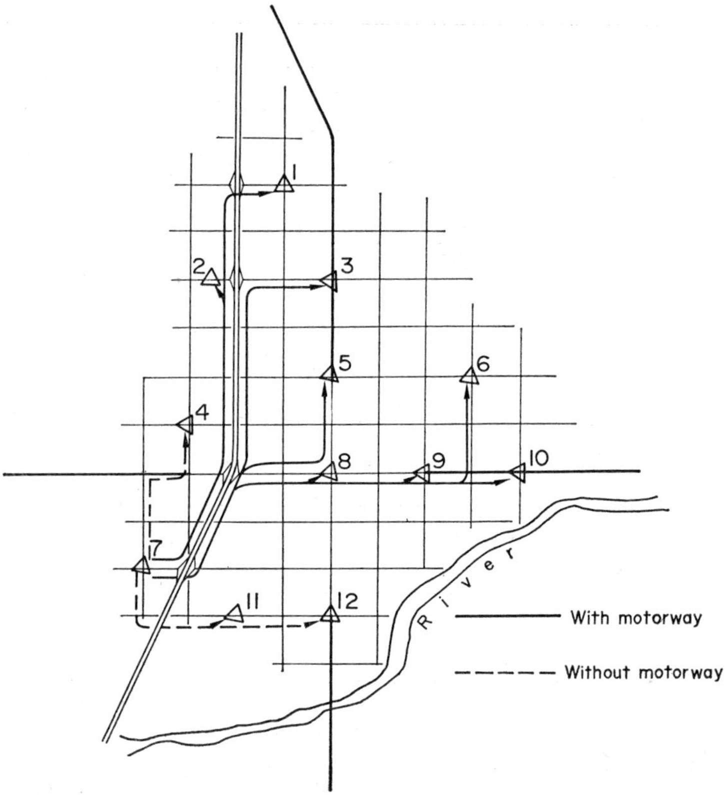


FIG. 1.7. Minimum path tree with origin at zone 7 centroid.

in the transport process. We provide now certain operational data relating to the transport network of ANZAC City in order that the interzonal transport impedances may be determined for use later on. The travel times on the road network are, of course, dependent on the level of flow on them. In this exercise we shall concern ourselves with peak-period conditions and assume that appropriate traffic observations have established the following peak-hour speeds:

(a) on the main roads 30 m.p.h.,
(b) all other surface streets 20 m.p.h.,
(c) on the motorway (estimated) 60 m.p.h.

It will be explained later how these assumed initial travel speeds may be adjusted as the planning process proceeds.

Other characteristics of the car transport situation that need specification relate to the terminal conditions, i.e. the time from parking point to final destination, viz.

| | |
|---|---|
| Zone 1 | 1 minute |
| Zone 6 | 3 minutes |
| Zone 8 | 5 minutes |

Operational data relevant to the public transport will be introduced later. In the meantime a table of interzonal travel times, with and without the motorway, is presented to fix our ideas on the manner in which transport impedance may be introduced into a practical problem. A good deal will be said in due course on the calculation of minimum paths through a network. The minimum times given in Table 1.3 may be readily checked by inspection. Figure 1.7 shows a set of minimum paths from zone 7 to all other zones.

## PLANS AND PLANNING

There is a need to distinguish between plans and planning. The former is generally synonymous with the solution but there is a strong tendency on the part of engineers and town planners to think of the solution as a physical plan—either drawn on paper or specified by dimensions and quantities. In the more general sense the plan may quite well be a strategy or policy to guide or govern a growth situation or a course of action.

Planning, on the other hand, is essentially the methodological process that leads to the plan. A principal purpose of this book is to study the methods and technique of land use and transport system planning. Planning may be undertaken in current time, i.e. an architect may plan an arrangement of rooms in a home to minimize the cost, entrap a maximum amount of sunshine, take best advantage of a view. However, in the land-use and transport context planning has become an extended operation in time. This is due in some measure to the fact that the period of gestation of major works and development and redevelopment schemes is measured in years and also to the fact that many ultimate goals—such as envisage the total elimination of slums or restoration of a complete harbor foreshore—would require decades for their realization. Whilst there is clearly a need for some balance between immediate implementation and long-term achievement, it would be fair to say that current planning practice more often than not reaches out for the crystal ball.

There are three basic methodologies of planning, viz.

(a) the projective (extrapolation),
(b) the deductive (synthesis),
(c) the objective (programming).

The art and science of planning is in the process of a complex evolution. The influence of operations research and systems analysis is causing the traditional projective approach to give way to deductive and/or objective planning procedures.

### *The Projective Planning Philosophy*

This philosophy has provided and to a large extent still provides the quantitative base for forward planning. It is essentially an open-ended extrapolation method. Data on the time trends of planning variables—traffic flows, vehicle ownership, residential densities—are projected forward with varying degrees of sophistication. One need not dwell on the inherent dangers in such a procedure when the projection interval is measured in tens of years. Further, the historical data base in many situations is of short duration. This state of affairs has led to a considerable exploitation of cross-section regression models which provide some link between the projective and the deductive planning philosophies.

Apart from the inaccuracy *per se*, the open-endedness encourages the planner to ignore the important effect of feedback. The dire predictions of the demographers on world population trends will surely be contained by the very consequences of their estimates. A more subtle disability arises from the geometric growth trend that is often exhibited in the short-term growth pattern of new developments. Figure 1.8 illustrates the basic characteristics of projective estimation.

It should be noted, however, that short-term extrapolation of the slope of a trend curve with subsequent adjustment is a correct and useful method of incremental planning or operations control.

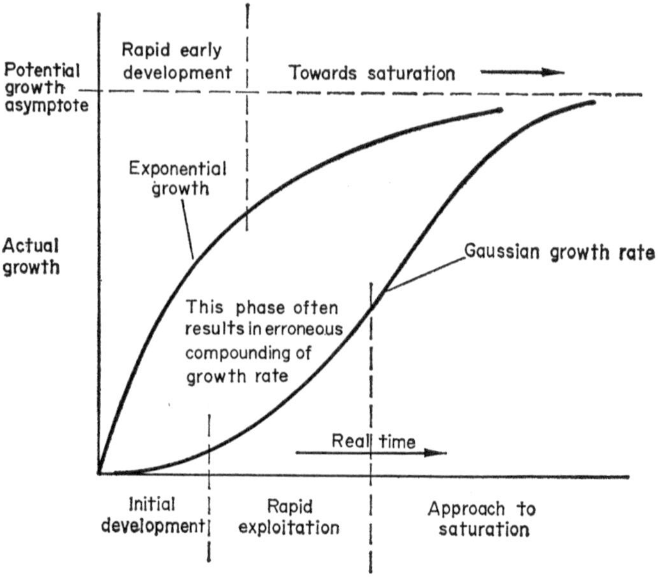


FIG. 1.8. Real time-growth functions showing the importance of saturation effect.

### *Deductive Planning*

The deductive planning philosophy envisages the synthesis of a future state of the system from laws, equations or models that govern its behaviour. This method is generally in vogue today for the transport component of the total land-use/transport system. The state of the art is, however, embryonic. Desire-line models are quite well developed but generation and assignment models (particularly intermodel assignment) require much more development. It is true also to remark that many of the more rational models are so "tightly" calibrated that they degenerate to regression or projective models.

The attempt to produce a theory of land-use and transport interaction has been characterized in its early stages by borrowing from other disciplines—physics, economics and probability theory. Gradually, however, a new jargon is developing around land-use and transport phenomena. Travel time and delay are new measures of transport impedance; traffic generation is a flux of attraction and traffic flow is analogous to current.

An important characteristic of the deductive planning process is that it takes account either implicitly or explicitly of feedback effects and this perhaps is the principal reason for its superiority over the open-ended projective planning philosophy.

### *Objective Planning*

The objective approach to planning is relatively new in the methodological sense but is perhaps the oldest conceptually. In objective planning the planner sets out to realize certain defined goals. This is in fact a pure planning philosophy. In the past the approach has been difficult to apply because of the multiplicity of solutions that will in general satisfy a set of end-point requirements. However, recent developments in mathematical programming methods have completely altered the prospects of the objective planning approach. A unique solution is possible by optimizing a valid performance criterion subject to a set of constraints that define the goals, specify the resources available (including the budget) and reflect the policies of the communities involved. For example, one may consider the distribution of travel desires between a set of origins and destinations that would minimize the total amount of travel involved subject to certain constraints relating to the number of trips that may originate or terminate at the respective origins and destinations. In other circumstances one may wish to maximize recreational trips in a given land-use context subject to capacity limitations on the road system.

### *Towards a Comprehensive Philosophy for Planning*

At this point in time there is no simple methodological recipe for the land-use and transport planner. Certain situations favour one or other of the approaches outlined above and efficacy of their application can be enhanced by having a clear specification of the object of the planning exercise—particularly in respect of time horizons and accuracy.

In the full context of extended urban or regional planning the planner should endeavour to partition the problem. A macro system with large land-use zones and major corridors of communication should provide the framework for subsequent detailed studies of residential subdivisions and industrial estates and their feeder and local transport facilities. The more sophisticated planning techniques have application to the large-scale problem. For example, mass transit systems may be competitively compared with new motorways in a main transport corridor instead of predetermining the public transport usage for the region as a whole before the major traffic demands are estimated.

Good tailoring of the methodology to the problem is an essential skill of the master planner. Certain styles demand special cloth. The projective approach is necessary for prediction of population trends, economic growth, vehicle ownership and other socio-economic indices. The uncertainties and other inherent difficulties of long-term physical plans or strategies are at least brought into the open by utilizing the objective planning method and the analysis of specific projects and operational activities such as by-passes, regional centres, transport terminals, one-way streets can be most effectively analysed by deductive planning techniques.

## THE PLANNING METHODOLOGY

Figure 1.9 is a flow diagram of the main computational phases of the deductive procedure for carrying out the planning task. The inputs are many and it is essential to distinguish one of them as the variable of analysis and regard the others as fixed. The question as to which one is selected depends on the purpose of the analysis. If a transport plan is the object then the "variable" input is a proposed transport system. The object could, of course, be to study the consequences of varying land-use plans, population levels, changes in economic activity. Having in mind that our principal concern is with transport planning, we have selected the transport system as the variable of analysis. The "fixed" inputs are made up of the present value inputs—existing population,

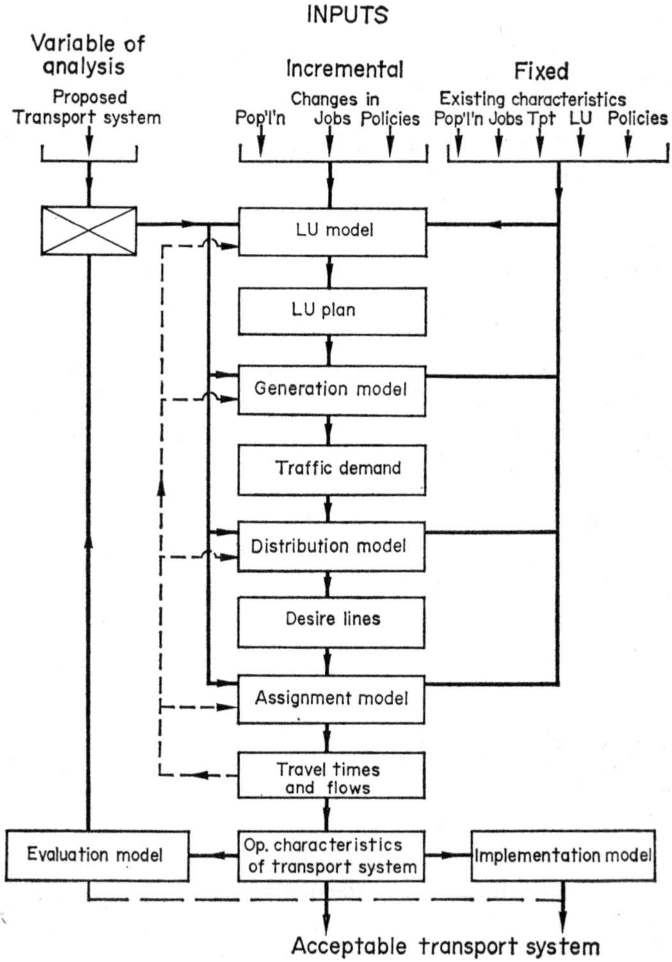


FIG. 1.9. Flow diagram of deductive planning process

jobs, land-use and transport relationships—and incremental inputs additional population, jobs, new zoning constraints, civic and transport policies. These incremental inputs are in general the result of projective planning methods. If incremental inputs are not included the process provides a methodology for a sensitivity analysis on the variable input.

There are six main analytical stages represented in the diagram by

The Land-use Model
Generation Model
Distribution Model
Assignment Model
Transport System Evaluation Model
Implementation Model

In a system as complex as that embracing the land-use and transport phenomena of a city there are many feedback channels. However, in outlining the fundamental properties of the systems approach we will do well to draw attention to the two principals. The inner loop—shown with the dotted line—determines the equilibrium state of the system in a predominantly physical sense. The travel times resulting from loading the travel demand onto the assumed transport system (the assignment model) are required to adjust the land-use plan, the traffic demand and the desire-line pattern. The system evaluation model—usually an economic model—is then used to assess the merit of the proposed transport innovations. Although the evaluation is influenced in certain respects by the implementation procedure it is necessary for practical reasons to reduce the feedback between evaluation and implementation to a minimum. A more comprehensive discussion will be given later on each of these key models in the transport planning process.

## CONCEPT PLANNING

It is desirable at this stage to introduce the notion of concept planning. This is necessary for practical as well as methodological reasons. Any real land-use transport system has an extremely complex infrastructure which governs many of the fixed inputs to the generalized planning process. However, some of these inputs may have upper and lower bounds and others may take on alternative discrete values—including zero. For example, it is in general essential to provide a geometrical definition of the zone boundaries and their centroids before proceeding with main planning process. But there are a number of different arrangements of the general disposition of the main land-use zones that should be assessed before fixing the geometry of the land-use pattern. The issue as to whether certain classes of industrial activity should be located within predominantly residential or segregated in an exclusively industrial context needs to be settled before detailed planning can start. In addition, guidance is needed for the specification of the initial transport solution—should, for example, a mass transit facility be included along with motorways and improved arterial roads. Concept planning can provide a framework of major boundary conditions and constraints.

Concept planning is carried out on a macroscopic scale—few land-use zones and major transport corridors. There is no need to set up detailed transport networks—in fact the transport system is implicit in the interzonal distance matrix.

There are two methodological approaches:

(a) comparison of a number of specific alternatives,
(b) optimization of an objective function subject to certain land-use and transport constraints.

In the former it is essential, of course, that the alternatives include amongst them a number of the good plans. There is, however, the possibility that a "best" plan is not conceived. The optimizing approach, however, automatically finds the best solution that is permitted by the boundary conditions and constraints.

## IMPLEMENTATION

In the field of land-use and transport planning the end product, i.e. the master plan or the development strategy, represents a target or framework for long-term action. Perhaps it would be proper for the planner to take his bow at this point and leave the stage. But even if he were to do this some other professional troupe must start up a new act and set about organizing the stage for the implementation of the grand design. Generally engineers and planners of the action agencies, the entrepreneurs and private developers, have the responsibility for preparing the detailed plans of the schemes, i.e. motorways, transport terminals, shopping centres, residential estates, which are implicit in the master plan. There is necessity however for a strategy to govern the implementation phase. Such a strategy emerges from economic and budget evaluations. This kind of economic analysis should not be confused with the evaluation phase of the master planning methodology even though it be based on economic measures and criteria.

Implementation may call for a master strategy or an optimization of a succession of programmes subject to more or less arbitrary budget constraints. In the former case the time scheduling and phasing of major works is the major problem to be solved. For example, such questions need to be answered as when should a second major airport be ready for operation, should the motorway to connect it to the city precede or follow it, should redevelopment of inner areas for high density have priority over the spread of the garden suburbs?

The capital budgeting problem, on the other hand, deals with a set of actual projects for which designs are available and for which costs are known and benefits may be estimated.

## THE LITERATURE

In concluding this introductory chapter it is appropriate to refer briefly to the literature. The quantity and rate of output of books, papers and articles on land-use and transport phenomena separately are huge. In parallel with the professional literature there is a great deal of operations research literature devoted to systems analysis and planning and an increasing amount of it is oriented towards land-use and transport phenomena—particularly the latter. If we add the voluminous reports of the many transportation studies that have been carried out in recent decades the task facing the student and the practitioner seeking knowledge and know-how would be formidable enough in terms of quantity alone. But the task is made much more difficult because of the lack of some kind of a common discipline with which to evaluate and assess it. The professional philosophies of the town planner and the transportation engineer stand far apart and their technical jargon is incommensurate. The practising traffic engineer and the traffic scientist have difficulty in communicating. The economist attempts to put common measures on both land-use and transport variables but has limited knowledge of their operational characteristics.

Though fragmented the literature contains many spectacular facets and reference will be made to them as appropriate, throughout this work. It must be admitted, however, that there is no consolidated work in book form that provides the student with some kind of unified philosophy of the subject of land-use and transport interaction and at the same time "talks" into it the new operations research based methodology of planning and systems engineering.

It is hoped that this book will make a small contribution to this dissideratum.

CHAPTER 2

# Land-use and Transport Interaction—Equations of State, Stability and Performance

## INTRODUCTION

It is common today to describe analytical relationships between the variables of a system as models. The origin of modelling and simulation dates back to the breadboard era of electronics and analogue computers when it was convenient to represent elaborate mechanical systems by simple and easily manipulable low-energy models. The rapid development in operational research during and since World War II and the application of analytical methods to large-scale operational and economic systems has led to an extension and abstraction of the model approach to mathematical representations of system behaviour generally. Queueing models, programming models, regression and functional models are the present-day equivalent of the formulae, equations and laws of mathematical physics. An equation of state is now a systems model.

But with all their sophistication there is something tentative and transitory about models—perhaps they capture the mood of society today! Although the jargon of modelling will appear quite extensively in this work, effort will be made to exercise some restraint with its use.

It is certainly true that knowledge today of the analytical and functional relationships between land-use and transport variables is transitory enough. But even at the risk of appearing somewhat pretentious the oneness of the land-use/transport context and the need to emphasize the notions of equilibrium and steady-state behaviour would seem to justify an attempt to set up equations of state for the system. In setting about this task one must seek a system of commensurate units in which to express the relationships between the land-use and transport variables. There are three possibilities—maybe more:

(a) land-use potential—transport impedance,
(b) economic demand—transport costs,
(c) travel desire probabilities—trip frequencies.

There is a certain degree of compatibility between all three. For our own purposes the potential/impedance concept will be the most instructive and we will direct our main attention to it.

## EQUATIONS OF STATE

Three equations suffice to describe the time space dynamics of a land-use transport system. They are:

(a) a land-use/transport interaction equation—ideally such an equation should express the amount of traffic and its spatial distribution as a function of the land-use transport system parameters;
(b) a transport impedance formula;
(c) an assignment equation for the distribution of the traffic between modes or within modes.

## THE LAND-USE/TRANSPORT INTERACTION EQUATION

It is clear that any land-use activity has a potential as an attractor or a source of persons or goods. At certain times a factory zone attracts workers and materials; the workers come from residential zones and the materials from mines and farms. At other times the residential zone becomes an attractor for workers returning home and for the products of industry. Office blocks, retail stores, entertainment and recreational centres all have a similar role as sources and sinks. Further, it is intuitively plausible that the larger the factory zone or the greater the number of dwelling units in a residential area the greater will be its attraction. It is plausible too that the ability to attract will diminish if it is more remote from the source of the persons or materials which it attracts. These propositions have, in a general way, been well substantiated by data on traffic generation and distribution patterns obtained from a variety of transport planning studies.

We have thus proposed that a zone of land-use activity has a potential, $L$ (say), to attract traffic from other land-use zones. We have proposed, too, that this attraction diminishes as the transport impedance between the zones concerned increases. We may say, therefore, the pull on a "unit" of land use, $o$, by an attracting land use separated from it by an impedance, $Z$, is proportional to

$$\frac{g(L_d)}{Z(t_{od})}.$$

If land use, $o$, is made up of $L_o$ units, then the traffic flow between the two zones may be written

$$T_{od} = K\frac{f(L_o)g(L_d)}{Z(t_{od})} \tag{2.1a}$$

where $K$ is the traffic generation "constant" which takes care of the dimensional adjustment and scaling of the arbitrary measures of land-use potential and transport impedance. Although there is need for research to be done on the nature of the functions $f$, $g$ and $Z$, there is nevertheless at this point in time a good deal of evidence available to validate some quite simple forms for these basic land-use and transport functions. *A priori* argument suggests $f$ and $g$ may be taken as linear functions of $L_o$ and $L_d$ respectively and $Z$ as an exponential function of $t$.

The law of interaction between land use and transport then takes on the familiar form

$$T_{od} = K\frac{L_o \times L_d}{t_{od}^n}. \tag{2.1b}$$

Although the $K$ of this "new" gravity model has none of the immutability of Newton's famous constant, it does, however, reveal the existence of a great deal of harmony in the firmament of man's socio-economic activity. $K$ is more appropriately styled the traffic generation parameter of a given land-use transport system. There will be more discussion about $K$ in Chapter 7 which is devoted to traffic generation. At this point it is sufficient to say that in most current applications of the gravity model the absolute level of traffic generation is determined independently and as a result $K$ may be eliminated and the gravity formula used as a distributor of a given number of predetermined trips. In what follows in this chapter it will be assumed that $K$ enters explicitly into the land-use/transport interactance formula as the determinant of either the total amount of travel or the actual number of trips.

## THE TRANSPORT IMPEDANCE EQUATION

We have already defined transport impedance and described its principal characteristics. The travel-time/flow relationship may be described analytically by the following formula:

$$t_{od} = \sum_{l} A(l, m, n)t_l \qquad (n \equiv od) \tag{2.2a}$$

where $A(l, m, n)$ is the land-use/transport incidence matrix, and

$$t_l = d_l t_l^0 \frac{1-(1-j_l)y_l}{(1-y_l)} \tag{2.2b}$$

where $d_l$ is the length of link

$$y_l = \frac{1}{s_l} \sum_{m} \sum_{n} A(l, m, n)q(m, n),$$

$s_l$ is the saturation flow of link $l$,

$q(m, n)$ is the component of the desire line ($od = n$) which takes the $m$th path between $o$ and $d$,

$j_l$ is the level of service factor of link $l$,

$t_l^0$ is the zero flow or "free speed" travel time on link $l$.

A full discussion of this travel-time/flow relationship is given in Chapter 3.

THE ASSIGNMENT EQUATION

The distribution of traffic amongst the transport facilities, i.e. amongst different modes or over alternative elements of a given mode (i.e. different paths through a road network) is generally described as inter- or intra-modal assignment; or alternatively modal split (i.e. amongst modes) or just assignment. Although there are many gaps in our knowledge of the mechanism of traffic assignment there is a great deal of evidence to support the idea that any shipper (including the one who ships himself) seeks to minimize the "cost" of his trip. At this point where our interest is directed towards a generalized description of the phenomenon of land-use and transport interaction in terms of psycho-physical variables we might assume that this is tantamount to minimizing the journey time and the shortest route would most certainly gain initial patronage. However, as the traffic load increases the travel time lengthens and some traffic will divert to the next shortest route if one exists.

Thus if a number of alternative routes are available traffic will distribute over them in such a way as to make the travel times equal. We can thus write

$$T_n = \sum_m q(m, n) \qquad n \equiv od, \tag{2.3a}$$

where $q(m, n)$ is that part of the traffic on desire line $n$ that uses route $m$ and

$$\sum_l A(l, 1, n)t_l = \sum_l A(l, 2, n)t_l \ldots n = 1, 2. \tag{2.3b}$$

Equations (2.1), (2.2) and (2.3) therefore may be quite fairly thought of as equations of state of the land-use/transport system. These equations of state will provide a basis for the development of more comprehensive land-use, generation, desire-line and assignment models. However, an example will serve to fix some of the basic ideas implicit in the heuristic discussion undertaken above and provide some kind of foundation for the notions of stability and steady-state performance of a land-use/ transport system.

AN EXAMPLE

Consider a simple land-use/transport system as shown in Fig. 2.1. Two land-use zones, $o$ and $d$, are connected by a road network comprised of a main arterial and an alternative path along a number of local streets. The two land-use zones may be taken as towns of populations 100,000 and 20,000 persons respectively. It may be assumed that surveys of the peak-hour travel between towns of this kind show that the land-use/transport interaction equation is of the form

$$T_{od} = K\frac{P_o P_d}{t_{od}}$$

where $T_{od}$ is the peak-hour travel,

$t_{od}$ is the travel time between $o$ and $d$,

$K$ is the traffic generation constant found here to be 60,

$P_o$ and $P_d$ is the population of $o$ and $d$ measured in units of 1000 persons.

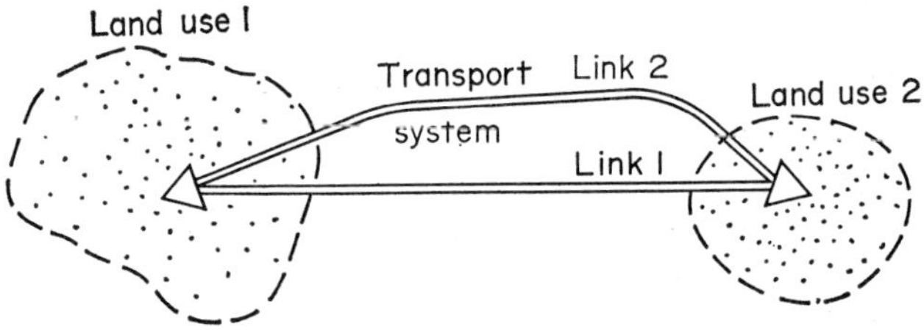


FIG. 2.1. Elementary land-use/transport system.

The simple linear function for $Z(t)$ is taken here to simplify otherwise tedious algebra. The transport system characteristics are summarized in Table 2.1.

The land-use/transport incidence matrix $A(l, m, n)$ may be constructed (Table 2.2).

Now

$$T_{od} = T_1 = K\frac{P_o P_d}{\sum_l A(l, m, 1)t_l}$$

$$= 60 \times 100 \times 20/\sum_l A(l, m, 1)t_l$$

$$= 12 \times 10^4/\sum_l A(l, m, 1)t_l \qquad m = 1, 2. \qquad \text{(i)}$$

Now for equal travel time assignment

$$\sum_{l=1}^{2} A(l, 1, 1)t_l = \sum_{l=1}^{2} A(l, 2, 1)t_l$$

or

$$t_1 = t_2 \tag{ii}$$

TABLE 2.1.

| Route | Length (miles) | Zero flow travel time-$t^0$ (min/mile) | Level of service factor-$j$ | Saturation flow (veh./hr) |
|---|---|---|---|---|
| 1 | 10 | 2·4 | 0·30 | 3000 |
| 2 | 12 | 3·0 | 1·00 | 2000 |

TABLE 2.2.

| Travel time | | | $t_1$ | $t_2$ |
|---|---|---|---|---|
| $q(m, n)$ | $n$ | $m$ \ $l$ | 1 | 2 |
| $q(1, 1)$ | 1 | 1 | 1 | |
| $q(2, 1)$ | 1 | 2 | | 1 |
| Saturation flow | | | 3000 | 2000 |

and

$$T_1 = \sum_{m=1}^{2} q(m, 1) = q(1, 1)+q(2, 1). \tag{iii}$$

The travel-time/flow relationship gives

$$t_1 = d_1 t_1^0 \frac{1-(1-j_1)y_1}{1-y_1}$$

where

$$y_1 = \frac{1}{s_1} \sum_{m=1}^{2} \sum_{n=1}^{1} A(l, m, n)q(m, n)$$

$$= \frac{q(1, 1)}{s_1}$$

and

$$y_2 = \frac{q(2, 1)}{s_2}.$$

Substituting values from the data table and expressing flows and saturation flows in 000's

$$t_1 = 10 \times 2{\cdot}4 \times \frac{1-0{\cdot}7q(1, 1)/3}{1-q(1, 1)/3}$$

$$= \frac{24(3-0{\cdot}7q(1, 1))}{3-q(1, 1)} \quad \text{(iv)}$$

also

$$t_2 = \frac{72}{2-q(2, 1)} \quad \text{(v)}$$

If we substitute (iv) and (v) in (ii) and simplify we obtain

$$1{\cdot}6q(1, 1)+3q(2, 1)+0{\cdot}7q(1, 1)q(2, 1)-3 = 0. \quad \text{(vi)}$$

Now (i), (ii) and (iii) give

$$q(1, 1)+q(2, 1) = 12 \times 10^4/t_1 \quad \text{or} \quad 12 \times 10^4/t_2 \quad \text{(vii)}$$

and substitution of (v) in (vii) yields

$$0{\cdot}6q(1, 1)+1{\cdot}6q(2, 1)-2 = 0 \quad \text{(viii)}$$

whence

$$q(1, 1) = 2{\cdot}20 \text{ (000's)} \quad \text{and} \quad q(2, 1) = 0{\cdot}43 \text{ (000's)}.$$

The total travel between the two towns is thus 2630 vehicles/hour and the travel time is approximately 45 minutes.

Although this example represents a highly simplified and perhaps artificial situation it demonstrates a number of fundamental points concerning what is commonly known as the transport planning process —or more correctly the land-use/transport planning process.

## STABILITY AND STEADY-STATE PERFORMANCE

Because of the extreme simplicity of the two town examples, an explicit solution was obtained by eliminating the travel time variable from the equations of state. However, the first four models of the planning process can be identified.

(a) The land-use model is implicit in the data which specified the populations of the two towns.

(b) The traffic generation model was conditioned by the value of $K$.

(c) The desire-line model was of course trivial—there being only one possible. Its actual value was of course determined by the travel time.

(d) The assignment was affected by balancing the travel times of the alternative routes between the two towns.

It is clear therefore that the travel time has a key role to play. In the illustrative example it can be manipulated algebraically to give a unique solution and in the more general planning methodology (Fig. 1.9) it is fed back in successive iterations until a balance is obtained.

It is not difficult then to perceive that a land-use/transport system is no different from any other physical or economic system and must in the long run settle down to an equilibrium state. It may of course suffer fluctuations in its behaviour that may have disturbing effects. In the short term where the land-use parameters are fixed, changes in the traffic inputs may cause transients in the transport system performance, common experience abounds with examples. On the first bright Sunday afternoon of spring, with memory dimmed of last year's record traffic jam, the family man sets out for the mountains or the beach only to find that this year's homeward trek is worse than ever before. For the remainder of the season his desire for Sunday outings is severely attenuated. The more experienced commuter, with the complications of getting to work always fresh in his mind, weighs up carefully the best route and mode, even perhaps the best place to work, and as a result plods his weary way through the transport maze of his urban abode in a surprisingly predictable manner. But unexpected changes in the characteristics of the system provide some spectacular examples of its inherent stability. In Australia in recent years a number of examples of this kind—transport strikes, accidents—have provided the road user with salutary lessons on this important point.

The temporary closure of a lane of the heavily loaded Sydney Harbour Bridge complex resulted in major delays on the first morning, but on the second morning the situation was virtually normal. The structural failure during the mid-morning of an important bridge in the inner Melbourne road complex caused chaos for homebound traffic the same day; but subsequently with the bridge out of action for inspection and repair there was no repeat performance, even though the system as a whole presumably worked at a lower level of efficiency. Transport strikes, introduction of new facilities—all provide striking evidence of the powerful feedback effect of traffic delays.

Variations in the land-use parameters generally result in responses with much longer time constants. It may be necessary to wait until the $(N+1)$th new city building turns out to be a white elephant before the developer is convinced that some decentralization is necessary. On the other hand, the football final and the bargain sale can produce some dramatic short-term transients.

Although much more research needs to be done to properly understand the complex mechanism of stability of the land-use/transport system, it is essential that the student should grapple firmly with the stability concept if he is to develop some maturity in his approach to the analysis and synthesis of land-use/transport systems.

## PERFORMANCE OF LAND-USE/TRANSPORT SYSTEM

If then a land-use/transport system settles into a state of equilibrium is there need for the planner and the engineer to concern himself further? Whilst there need be little concern that the transport as such will grind to a halt, it may be extremely inefficient and the same may be said for the system as a whole. Further, there is the matter of growth to be considered.

The transport system engineers and the land-use planners have two responsibilities:

(a) to improve the efficiency of the system,

(b) to plan for its growth and development.

At this point we will raise in a general way the question of describing the equilibrium or steady-state performance of the system. How do we measure its efficiency?

These are difficult questions but answers must be found. The basic problem is one of criteria and the problem really becomes one of establishing a hierarchy of criteria that have consistency and compatibility. The obvious starting point is to devise a land-use and transport system that makes the greatest contribution to the overall welfare of the community it serves. There is a very wide range of activities that makes up the socio-economic activity of a community, but to generalize broadly the prosecution of these activities calls for work, marketing and recreation. Although the process of being in transit provides some satisfaction in itself, it would in general be true to say that it is a charge against the activities in question. It is clear therefore that it should be our more limited aim to minimize time spent in or cost of transport either by making the transport system more efficient or planning the spacial relationship of the land uses so as to minimize the amount of travel.

We shall see later that there is a great deal of compatibility between exogenous criteria that seek to minimize transport costs and the conceptual role of travel time in the equations of state.

Although the minimization of travel time or cost is a generally adequate objective in land-use and transport planning, there is one major difficulty. Our land-use/transport interaction equation shows that as the travel time is reduced the number of trips will rise. In other words, the provision of a new transport facility may well result in a greater amount of travel activity than previously existed. But in so far as more travel in this sense means more trips, then presumably more community desires are being satisfied. The situation is closely analogous with the economic theory of demand and the associated concept of consumer surplus.

Figure 2.2 shows the familiar relationship between price and demand and cost and supply for a given economic commodity. For a given demand and supply situation the point of intersection, $P_1$ (say), establishes the market price of the product. The consumer surplus is the area (unit price × amount) above $C_1P_1$. If now the efficiency of

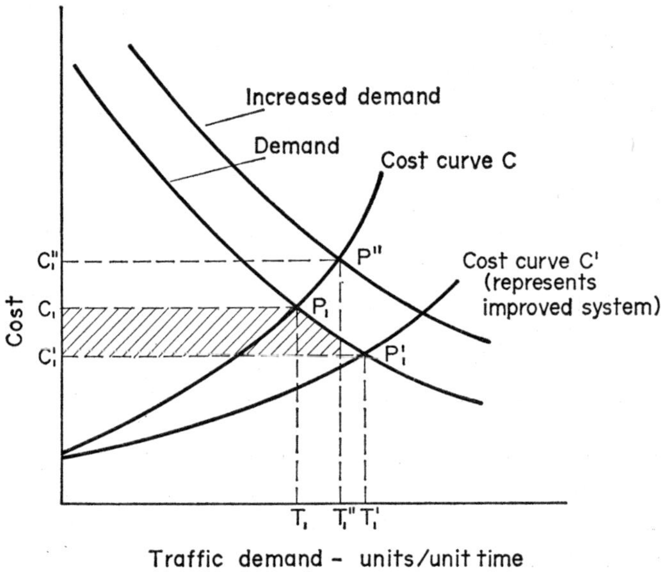


FIG. 2.2. Consumer surplus notion of benefits.

production is improved, resulting in the curve $C'$, then price/demand equilibrium is established at $P_1'$. The net surplus, $W$, is thus represented by the area $C_1P_1P_1'C_1'$, i.e.

$$\tfrac{1}{2}(T_1+T_1')(C_1-C_1').$$

If, on the other hand, the demand characteristic changes to $d'$ and the supply characteristic remains unchanged, then again there is a net change in the consumer surplus. Provided the new demand curve is sensibly parallel to the original, then the areas of the consumer surplus are proportional to the square of a characteristic dimension, say $T$. The net difference in the areas is thus proportional to

$$(T_1''^2 - T_1^2) = (T_1 + T_1'') \cdot (T_1'' - T_1)$$

but $(T_1'' - T_1)$ is itself approximately proportional to $(C_1 - C_1'')$. Therefore

$$W = K(T_1 + T_1'')(C_1 - C_1'').$$

The student may now ponder on the striking similarity of these arbitrarily drawn economic characteristics and those resulting from plotting the first and second equations of state. The gravity model form of the land-use/transport interactance law is essentially a demand curve and the cost characteristic has the form of a travel time/flow relationship. It is of some philosophical interest that whilst the demand curve of the economist usually shows a finite value for the product at zero demand, the gravity type interactance formula is asymptotic to the travel time (i.e. cost) ordinate. Although this state of affairs can be readily reconciled by using a modified travel time or cost function, such as that of Tanner,[(3)] it is perhaps of some significance that in the transport context the price of a first trip is near enough to infinite if we consider the first trip to the moon or the cost in human life on many pioneering journeys of exploration.

## OTHER ANALOGUES

It was indicated at the beginning of this chapter that economic and probabilistic analogues of the land-use/transport system may be developed. Generally they add little to the basic concepts illustrated above in terms of the psycho-physical variables of attraction and travel time impedance.

### *Economic Models*

An economic model of the land-use/transport system would replace the traffic potential concept with a more generalized demand function and the travel time function with a cost/flow relationship. The latter is straightforward enough but the former presents considerable difficulty. However, if little is to be gained in resorting to a general theory based on economic variables, economic concepts have a vital place in the detailed methodology of planning. Many of the parameters of the various traffic generation functions are economic ones; in intermodal assignment relative costs are more discriminating than travel times. In evaluating the performance of both the land-use/transport system as a whole and the transport system in particular, economic criteria are pertinent. Implementation is essentially a matter of economics. We shall make reference to various economic models as the work proceeds.

### *Probability Models*

There has been a considerable exploitation of probability concepts in many aspects of land-use and transport planning. Probability models for land-use and desire-line distribution have been developed and certain assignment problems have been studied with probability distributions. In so far as queueing theory provides a basis for the planning and operational control of transport terminals, probability methods are of extreme importance and utility.

The competing and intervening opportunities' models of Stouffer and Schneider have been applied to the location and desire-line problems and thus provide a basis for a theory of land-use and transport interaction. The opportunity models do not introduce transport-system impedance explicitly and as a consequence are somewhat less adaptable for introducing the basic concepts of interaction.

Transport impedance is, of course, taken account of implicitly by ranking the land-use zones in terms of distance, travel time or cost. In certain applications this procedure is advantageous for it eliminates the need to develop a formal impedance function. We shall discuss the opportunity models in the chapter dealing with desire-line patterns.

## THE OBJECTIVE FORMULATION OF THE LAND-USE/TRANSPORT SYSTEM EQUATIONS

If we consider a set of $o$ origins and $d$ destinations, the traffic activity of the system may be described by the land-use/transport incidence matrix as given in Chapter 1. In the example given a set of $n$ desire lines was arbitrarily specified. However, the land-use/transport interaction concept implies that the transport system itself will influence both the level of land-use activity and the traffic desire-line pattern. We can, however, express the interdependence of the transport and land use with the following system of constraints:

$$\left.\begin{aligned}\sum_{o=1}^{O} T_{od} &\geqslant E_d^P\\ \sum_{o=1}^{O} T_{od} &\leqslant E_d^F\end{aligned}\right\}\ d = 1, 2, \ldots, 0,$$

$$\left.\begin{aligned}\sum_{d=1}^{D} T_{od} &\geqslant R_o^P\\ \sum_{d=1}^{D} T_{od} &\leqslant R_o^F\end{aligned}\right\}\ o = 1, 2, \ldots, 0, \qquad (2.4)$$

where $E^P$, $E^F$ and $R^P$, $R^F$ are present and future jobs and populations. The traffic generation model must be expressed with a further constraint

$$\sum_{o} \sum_{d} T_{od} = \sum E \text{ (the total future work force).} \qquad (2.5)$$

However, if these constraints are to yield a unique solution it is necessary to maximize or minimize some functions of the traffic activity, i.e. the $T_{od}$'s. This can quite clearly be the total amount of travel which has already been defined as

$$\sum_{l}\sum_{m}\sum_{n} A(l, m, n)t(l)q(m, n). \qquad (2.6)$$

This definition of travel function implies that the travel desires have been allocated to a real traffic network and as a result the capacity constraints of the network elements must be recognized, viz.

$$\sum_{m}\sum_{n} A(l, m, n)q(m, n) \leqslant s_l \qquad l = 1, 2, \ldots. \qquad (2.7a)$$

The relationship between the flows on the paths and the traffic demand (or desire) lines requires specification

$$T_{od} = T_n \qquad n = 1, 2, \ldots, O \times D$$

and (2.7b)

$$\sum_{m} q(m \,.\, n) = T_n.$$

This method of describing the land-use/transport system is particularly flexible so far as its spatial characteristics are concerned and the L.P. formulation permits the mathematical optimization to take account of many variables and constraints. However, a price must be paid—to wit the inability to take account of the flow dependence of travel time.

The effect of flow on travel time in the links can, of course, be introduced into the analysis by a succession of iterations after adjustment of link travel times in the light of the previous solution. However, it is as well to say that the objective planning approach is best suited for the longer term and concept planning operations and in such circumstances link lengths and somewhat arbitrarily chosen travel speeds provide a sound basis for such plans.

It is appropriate also to point out that the above formulation of the land-use/transport system assumes that the total number of trips is given. Here again an iterative process may be employed to adjust the trip generation. Further, if concept planning remains the principal role then the primary definition of the land-use parameters is best directed at future population and job totals and the resultant work trip traffic pattern. In this situation induced trips are of minor consequence.

In this approach the evaluation of the equilibrium state of the system is implicit in the optimization of the objective function. Because the constraint equations (2.4) permit the levels of land-use activity in the origin and destination zones to change and consequently the desire-line pattern as well. Somewhat paradoxically, it is the transport network that provides the most severe constraints—the capacity constraints. However, the alternative paths through transport networks permit considerable opportunity for optimization. In other words, the form of the objective function given in (2.6) allows a good measure of optimization of the total land-use/transport system. This optimization process corresponds largely to the inner travel time feedback loop of the conventional planning process model in Fig. 1.9 for it is essentially an optimization of the operational or user costs of the system. The outer feedback loop involving the evaluation model takes account of the total cost optimization. For example, the balanced (or optimum) operational situation may result in some components of the proposed transport network being underloaded. Whilst this is operationally satisfactory perhaps the benefits from such a situation may not match the capital cost of providing it. The reverse case may also be equally unsatisfactory. An element of the transport network may be working so close to saturation that its extra operational costs would more than cover a larger initial investment.

An objective function which included capital as well as operational costs could be written down, but at this stage a good deal more research and experience is needed to build the many parameters of a real system into such a model. In the meantime the programming or objective approach can be applied separately to the evaluation phase. This can be done by applying various cost/effectiveness and capital budgeting analyses.

In concluding this section it is appropriate to refer again to the problem of induced or generated demand resulting from the improvements themselves. It would seem that in such circumstances we must attribute an intrinsic value to the trip itself. If this is done the net profit or worth of the logistical activity of the land-use/transport system may be written

$$W = \sum_n T(n)V(n) - \sum_l \sum_m \sum_n A(l, m, n)q(m, n)\{C(l)+a(l)\}$$

where $W$ is the "operational" profit or worth of the system,

$V(n)$ is the average value of trips of desire line $T(n)$,

$C(l)$ and $a(l)$ are the operational and accident cost rates for link,

and the other symbols are as previously defined.

It is of some relevance to observe that the second term of this expression incorporates the traditional objectives of the various professional bodies and agencies concerned with transport—"promotion of the efficient and safe movement of persons and goods".

## EQUATIONS OF STATE AND PLANNING PHILOSOPHY

In rounding off this attempt to formalize the intuitively obvious interdependence of the land-use and transport components of the total system, it should be pointed out that a great deal of technique and "know how" must be acquired in applying the so-called equations of state to particular problems. At the level of metropolitan planning the land-use/transport system must be defined in grossly macroscopic scale and the equations or models should be used as creative planning tools. To calibrate laws that constrain human activity with data derived from present behaviour is bound to straitjacket future opportunities. On the other hand, the interaction between a proposed supermarket and the traffic on the road it fronts would call for some detailed knowledge of the factors that influence the land-use and transport variables and parameters.

Although deterministic equations of the land-use/transport system provide the basis for the deductive planning approach, it should be mentioned that the definition of the variables and parameters and the rationale of the equations guide the formulation of the objective functions in the objective planning procedures. For example, the minimization of passenger minutes of travel in optimizing a desire-line pattern is in harmony with inverse travel-time function of the gravity model formulation of the same problem.

CHAPTER 3

# Transport Impedance and Traffic Flow

## INTRODUCTION

We have already made introductory comment on the nature of transport impedances. A vast array of experimental data on all kinds of traffic-flow situations (ticket windows, intersections, airport runways, lengths of highway, and so on) exhibits a performance characteristic of the type shown in Fig. 1.6. Remembering that a unique feature of the traffic of physical communication is its "molecular" or discrete and in general stochastic nature, it is appropriate to resort to the elegant theory of queueing for the theoretical foundations of the impedance relationship.

The basic theory of queueing is given in statistical texts and there are now a number of very useful books on the subject.[4–6] There is also an enormous literature which until recently was oriented primarily in the direction of telephone traffic. In the last decade, however, many transport applications have been presented.[7–16] It is not proposed to intrude here on this splendid literature. Though it be elegant, the rigorous mathematics of queueing is, however, very sophisticated and the student needs some firm guidance if he is to develop a good "feel" for the power of this tool of traffic analysis.

## CLASSIFICATION OF QUEUEING SYSTEMS

Before describing the usual classification pattern mention should be made of the fact that the mathematics of queueing has been restricted to the treatment of what may be best described as *barrier*-type queueing elements. By this is meant that the queueing facility, be it a toll booth, an intersection, a retailing point, a parking stall, is discretely located in space. However, of considerable importance to the traffic analyst is the case of the *continuously distributed* traffic element, notably a finite length of roadway, rail track or footpath. The travel time/flow characteristic of traffic on such an element is of the same type as that for a barrier element, but at this point in time its form can only be established by analogy and its parameters by experiment.

In the more formal context then of barrier queueing it is instructive to adopt a $2 \times 2$ classification in distinguishing between the nature of the traffic loading on the one hand and method of interconnection on the other. The traffic loading may be either stochastic or deterministic and their interconnection in series, parallel or some series/parallel combination.

It is not intended to develop the subject of queueing theory in a formal mathematical manner. The purpose of this chapter is to review and explain some of the more useful formulae and illustrate their use with examples.

## BASIC QUEUEING SITUATIONS

There are a number of simple queueing situations which individually or in some combination provide a basis for analysing a very wide range of transport problems. They are:

(a) the single channel—toll barriers, intersection approaches, kerbside bus stops;

(b) multi-channels in simple parallel arrangement—parking lots, shipping berths, railway stations, loading docks, bus terminals, ticket windows, baggage counters;

(c) queueing for gaps in a priority traffic stream—minor road junctions, ingress and egress to land uses fronting main roads and arterials, access problems generally, aircraft departures;

(d) time sharing of right of way—traffic signals, pedestrian signals, railway level crossings;

(e) distributed traffic flow elements.

THE QUEUEING PROCESS—DEFINITIONS AND TERMS

Figures 3.1(a), (b) and (c) illustrate diagrammatically the principal functional components of a queueing system for single and multiple service facilities.

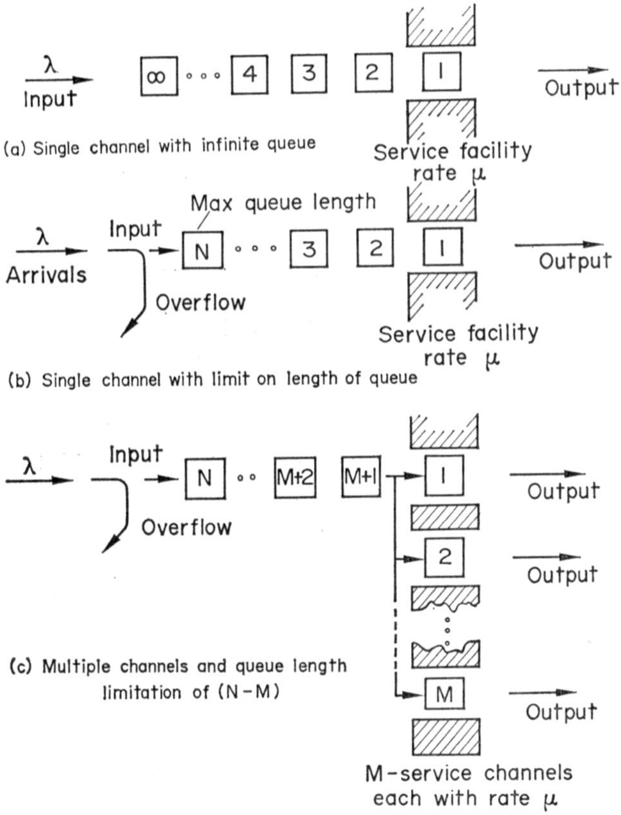


FIG. 3.1. Diagrammatic representation of queueing system arrangements.

### *Service Times*

The service time is the time to complete one full cycle of the operation appropriate to the type of service being sought. It is not always easy to measure, but in most cases it is equal to the time between successive departures from the service facility *provided a queue exists.* In general, each service time is different and a sufficient sample of individual service times can be used to construct a service-time frequency curve—Fig. 3.2(a). It is more common to construct the cumulative distribution function, Fig. 3.2(b). This is usually known as the service-time distribution. Mostly it is sufficient to describe the service-time distribution in

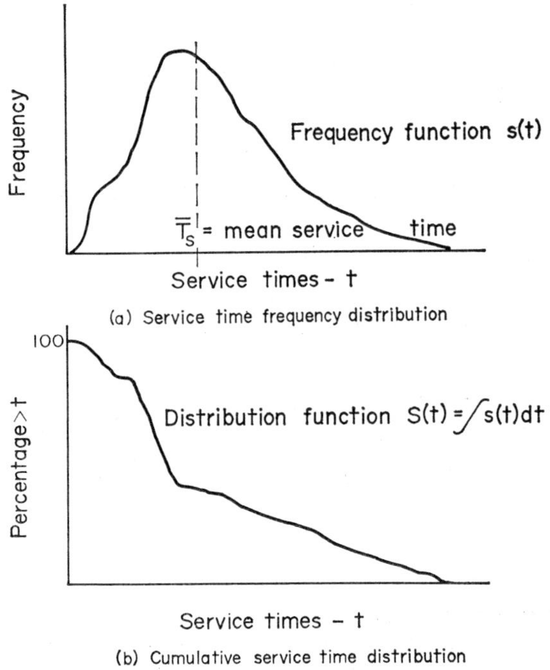


FIG. 3.2. Service-time distributions.

the usual way by calculating its mean and variance $\overline{T}_s$ and $\sigma_s^2$ respectively. $\overline{T}_s$ is known as the mean service time and its reciprocal, $\mu$, the mean service rate. It is synonymous with the saturation flow rate or the capacity of the traffic element.

$$\mu \text{ (mean service rate)} = \frac{1}{\overline{T}_s} \text{ per second,}$$

$$s \text{ (saturation flow)} = 3600\mu \text{ per hour.}$$

Service-time distributions may be of arbitrary form, but the considerable amount of experimental data that has been accumulated during recent years indicates that such distributions have well-defined general characteristics. Service times of an "unrehearsed kind" subject to a range of independent factors tend to have an exponential distribution, e.g. times taken at petrol stations. On the other hand, practiced tasks such as starting up at an intersection, paying tolls at a toll barrier exhibit a great deal of regularity. A family of distributions of the gamma type known as Erlang Service Time Distributions—after the Danish pioneer of the application of queueing theory to traffic (albeit telephone traffic)—provide a useful set of theoretical forms to which to fit experimental data.

The Erlang distributions characterized by the Erlang number, $K$, may be written

$$S(t) = e^{-K\mu t} \sum_{n=1}^{K-1} \frac{(K\mu t)^n}{n!}, \tag{3.1}$$

when $K = 1$ $\quad S(t) = e^{-\mu t}$ (i.e. the exponential distribution), (3.1a)

$K = \infty$ $\quad S(t) = \overline{T}_s$ for $t < \overline{T}_s$,

$= 0$ for $t > \overline{T}_s$ (i.e. a constant distribution), (3.1b)

and $K$ is the reciprocal of the coefficient of variation, i.e. $\overline{T}_s^2/\sigma_s^2$.

Figure 3.3 gives a graphical representation of a typical family of Erlang distributions. Figure 3.4(a), (b), (c), (d) shows some typical service-time distributions that have been obtained by students of the author in Sydney and California.

### *The Traffic Input to a Service Facility*

The traffic input is defined in terms of a distribution of intervals of time between the units seeking to use the service facility. The time intervals or headways are measured up-stream of the rear end of the queue (if any) that has formed at the service facility.

As with the service-time distribution the arrival distribution may have an arbitrary form, but in a wide range of situations arrival distributions can be approximated by one of the Erlang distributions. It is of interest and relevance to note, however, that whereas service times tend to the higher Erlang numbers (i.e. high $K$) the arrival distributions have $K$ values in the lower range. Because of the intractability of theoretical queueing formulae for other than exponential or regular distributions this fact is of some considerable practical value. Good approximate solutions to real problems can be made by assuming

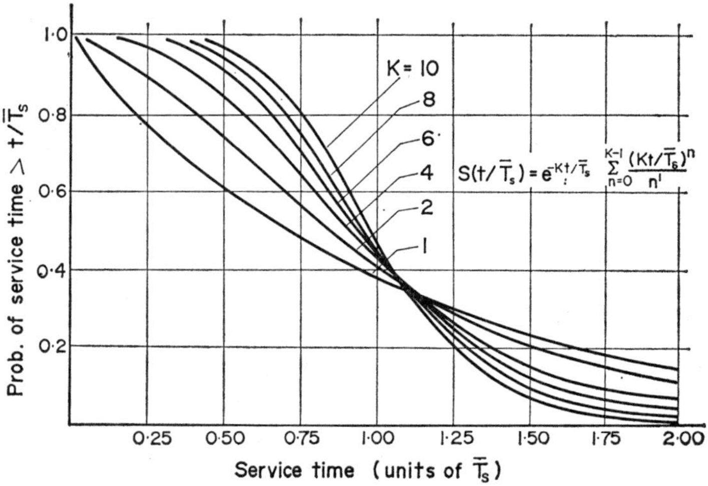


FIG. 3.3. Family of Erlang service-time distributions.

regular service and exponential arrivals. It will be seen later so far as the single-channel queueing situation is concerned, there is little difficulty in dealing with arbitrary service characteristics provided the arrival stream is exponential.

If the average headway between arrivals is $\overline{T}_a$ sec the average arrival rate is

$$\lambda = 1/\overline{T}_a \text{ units/sec and } q = 3600\lambda$$

where $q$ is the traffic flow in vehicles/hour.

So far as highway traffic flow is concerned there has been considerable effort in recent years devoted to the study and description of inter-vehicle headways. Comment on this important phenomena should, however, be prefaced by reference to the remarkable work of Adams, who in the mid-thirties demonstrated that it could be represented by a simple exponential distribution. Although many subsequent experiments and studies show that the headway structure of the traffic stream is anything but simple, Buckley[17] has presented the most

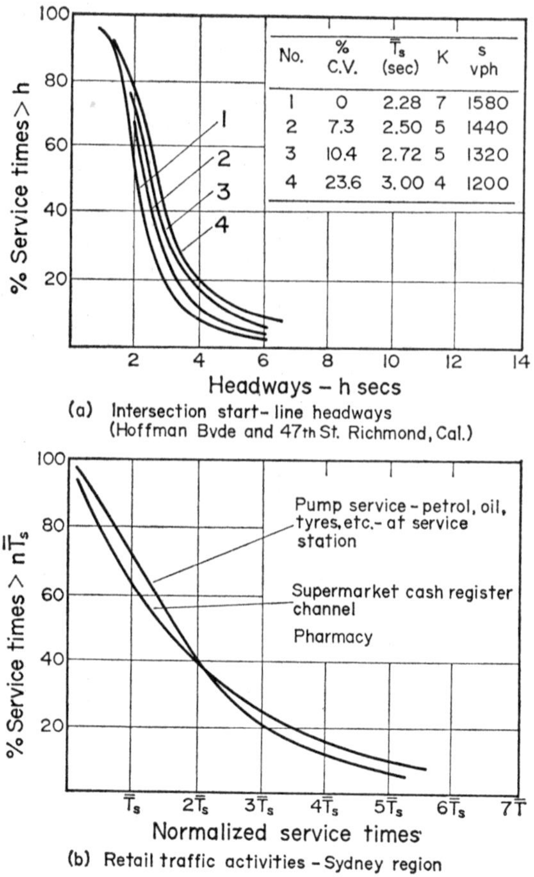


FIG. 3.4 (a) and (b). Service-time distributions for various types of traffic activity.

conceptually adequate model of its essential character. He has compounded the headways from two distributions—the one representing those vehicles that are, because of their proximity to the one in front, deterministically tracking it and the other the remaining vehicles.

For the exponential arrival distribution

$$A(t) = e^{-\lambda t} \tag{3.2a}$$

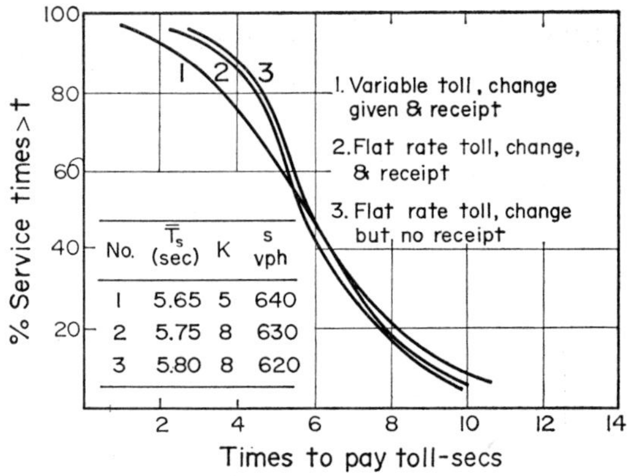


(c) Distribution of toll service times on Sydney harbour bridge for various operational procedures

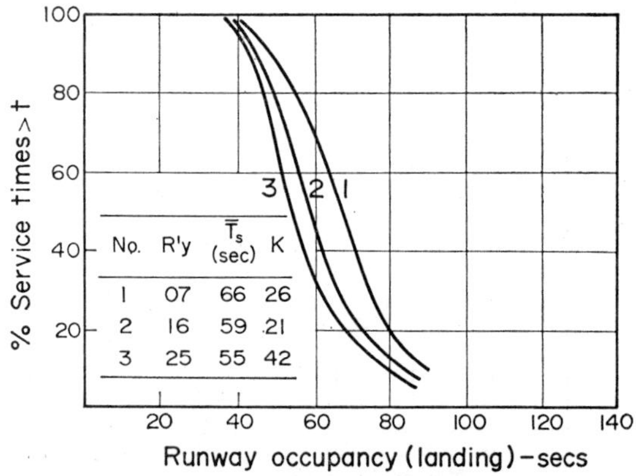


(d) Service distributions of runway occupancy times at Sydney Kingsford Smith airport

FIG. 3.4 (c) and (d). Service-time distributions for various types of traffic activity.

and for Buckley's formulation the frequency function is

$$a(t) = \phi g(t) + (1-\phi)h(t) \tag{3.2b}$$

where $g(t)$ is the frequency of follow-up headways and $h(t)$ the frequency of the others. $\phi$ is the proportion of follow-up headways.

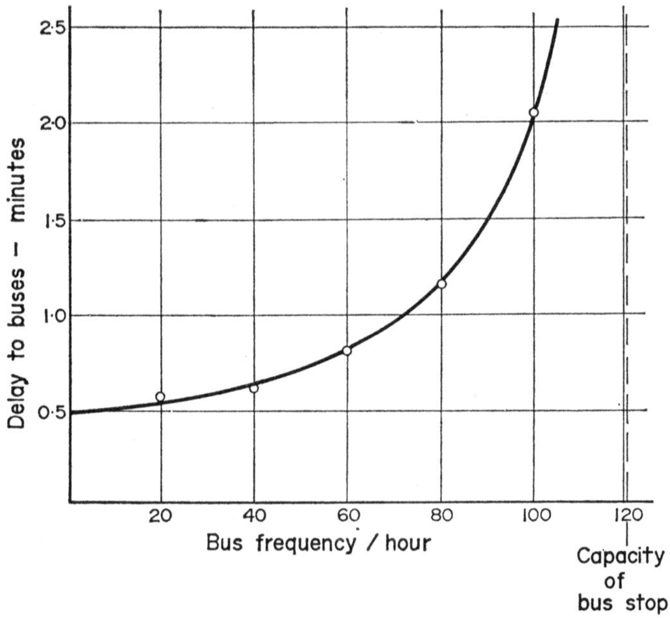


FIG. 3.4(e). Total delay (i.e. loading time plus waiting) suffered by buses at kerbside stop.

### *Traffic Intensity, Degree of Saturation, Utilization Factor*

The ratio of the average arrival rate to the average service rate is variously known as the traffic intensity, degree of saturation, utilization factor.

$\frac{\lambda}{\mu}$ is usually designated by $\rho$ and $\frac{q}{s}$ by $y$.

One might observe that the U.S. practice is to use $V$ (i.e. volume) for arrival flow and $C$ (i.e. capacity) for the saturation flow.

In multi-channel systems it is common to use the term traffic load. The traffic load $A$ is given by

$$A = q\overline{T}_s = M\rho.$$

Note: $\overline{T}_s$ is the mean service time in individual channels.

### *The Queue*

The queue is the line of units that are delayed behind the one or ones receiving service. A variety of measures are used to define it—its length, the time units are delayed in it, the probability that it exists at all or exceeds a certain size. It is a stochastic variable and the above measures are usually specified by means and variances of the distributions of the measures mentioned above.

Closely associated with these measures of the queue is a corresponding set relating to the system, i.e. the number in the system, the delay in the system. These measures include the unit or units in service as well as those in the queue or waiting line.

### *Queue Discipline*

This term is used to describe the method of working or managing the queue. There are a number of disciplines, viz.

First come first served.
Diversion from the system when the queue reaches a certain length—including zero length.
Diversion after a certain time.
Diversion proportional to queue length or waiting time.
Priority components.
Single queue or separate queues (multiple channel system).

## SOME IMPORTANT QUEUEING RESULTS

The rigorous treatment of the single-channel queueing facility is of great consequence in the development of the theory of traffic flow. In the area of planning and transport-system analysis, the multiple-channel queueing model has more general application. Further, the simple results for the single-channel system may be obtained as special cases of the multiple-channel system. Although the results given below are standard in a number of works on queueing theory, the idiom of their expression here has been inspired by the approach taken by Rallis[18] in his most valuable treatise on capacity of transport centres.

### *Multiple-channel Queueing System with M-parallel Channels each having a Mean Service Time $\overline{T}_s$, of Exponential Form*

If the demand on the system (i.e. the arrival rate) is $q$ and units are turned away when the number in the system (i.e. in service and queueing) reaches $N$, then the steady-state probability that there will be $n$ in the system is

$$P_n = \frac{A^n}{n!} P_0 \qquad 0 < n \leqslant M$$

or

$$= \frac{A^M}{M!}\left(\frac{A}{M}\right)^{n-M} P_0 \qquad M \leqslant n \leqslant N \tag{3.3}$$

where $A = q \, . \, \overline{T}_s$.

Although it is not intended to give mathematical proofs of the queueing results developed in this chapter the edifice to be erected on the above formula is such as to warrant a formal proof and this is given in the appendix to this chapter.

It should be noted that for $n \leqslant M$ there is no queue and for $n > M$ the queue has length $(n-M)$ units at any instant.

The value of $P_0$ is readily obtained from the boundary conditions of the situation being considered.

Let us consider first the probability that the system is being fully utilized. Clearly this is so when $n = M$, thus

$$P_M = (A^M/M!)P_0.$$

If all units arriving to find the system full are turned away then at no time will the number in the system exceed $M$, i.e. the only possible states of the system are $n = 0, 1, 2, \ldots, M$.

Thus

$$P_0+P_1+P_2+\ldots P_M = 1,$$

$$P_0+AP_0+\frac{A^2}{2!}P_0+\ldots \frac{A^M}{M!}P_0 = 1$$

and

$$P_0 = 1/(1+A+A^2/2!+A^M/M!),$$

hence

$$P_M = \frac{A^M/M!}{1+A+A^2/2!+\ldots A^M/M!}. \tag{3.4}$$

This is known as the probability of rejection or probability of loss formula and we will designate it with the symbol $P_L$. Not only is it an important measure of performance but it is a most important mathematical function in the queueing theory context as many other "clumsy" formulae are simplified by its use. It is the ratio of the Poisson to the Cumulative Poisson Function. It is difficult to compute and although the standard books of tables have tabulations of the numerator and denominator, this simple relationship between them is worthy of tabulation in its own right. Tables 3.1 (a) and (b) develop such a tabulation in two stages and it will be found of genuine utility to the student.

Although $M$ values as high as 100 are adequate for most situations occurring at terminals, it will be shown later that parking areas may be regarded as multi-channel queueing facilities. In such cases use may be made of Stirling's formula for evaluating $P_L$.

The probability of rejection is a useful measure of performance and one much used on telephone traffic systems. It certainly is a measure of the level of service offered by the facility in question. However, in telephone traffic problems the authority responsible for the facilities does not need to worry about the rejected units, i.e. the telephone callers—they mutter, put down the handset and go about other business until they decide to dial again. However, in physical traffic systems the rejected units, e.g. cars, lorries, ships, aircraft, must "cruise around" or wait in the queue. Other measures of effectiveness need to be investigated.

TABLE 3.1. PROBABILITY OF REJECTION—$P_L$%[18]

(a)

| Traffic load | Number of channels | | | |
|---|---|---|---|---|
| $A$ | $M = 1$ | 2 | 3 | 4 |
| 0·1 | 9 | 0·5 | 0 | 0 |
| 0·2 | 17 | 2 | 0·1 | 0 |
| 0·3 | 23 | 3 | 0·3 | 0 |
| 0·4 | 29 | 5 | 0·7 | 0 |
| 0·5 | 33 | 8 | 1 | 0·2 |
| 0·6 | 38 | 10 | 2 | 0·3 |
| 0·7 | 41 | 13 | 3 | 0·5 |
| 0·8 | 44 | 15 | 4 | 0·8 |
| 0·9 | 47 | 18 | 5 | 1 |
| 1·0 | 50 | 20 | 6 | 2 |
| 2·0 | 67 | 40 | 21 | 10 |
| 3·0 | 75 | 53 | 34 | 21 |
| 4·0 | 80 | 62 | 45 | 31 |

(b)

| Traffic load | Number of channels | | | | | | | |
|---|---|---|---|---|---|---|---|---|
| $A$ | $M = 1$ | 5 | 10 | 20 | 30 | 40 | 50 | 100 |
| 1 | 50 | 0·3 | 0 | 0 | 0 | 0 | 0 | 0 |
| 2 | 67 | 4 | 0 | 0 | 0 | 0 | 0 | 0 |
| 3 | 75 | 11 | 0 | 0 | 0 | 0 | 0 | 0 |
| 4 | 80 | 20 | 0·5 | 0 | 0 | 0 | 0 | 0 |
| 5 | 83 | 28 | 2 | 0 | 0 | 0 | 0 | 0 |
| 10 | 91 | 56 | 21 | 0·2 | 0 | 0 | 0 | 0 |
| 20 | 95 | 76 | 54 | 16 | 0·8 | 0 | 0 | 0 |
| 30 | 97 | 84 | 68 | 38 | 13 | 1 | 0 | 0 |
| 40 | 98 | 87 | 76 | 52 | 30 | 12 | 2 | 0 |
| 50 | 98 | 90 | 80 | 61 | 42 | 25 | 10 | 0 |
| 100 | 99 | 95 | 90 | 80 | 70 | 61 | 51 | 8 |

***The Probability of an Arriving Unit Being Delayed***

We distinguish between finite and infinite values for $N$.

(a) $N$ finite and greater than $M$.

The probability of being delayed is the sum of the independent probabilities that an arriving unit will find either $M$, $M+1$, ..., $N$ units already in the system. Thus the probability of delay $P_D$ is

$$P_M + P_{M+1} + \dots P_N = \frac{A^M}{M!} + \frac{A^M}{M!}\left(\frac{A}{M}\right) + \dots \frac{A^M}{M!}\left(\frac{A}{M}\right)^{(N-M)} P_0$$

$$= \frac{A^M}{M!}\frac{M}{M-A}\left\{1-\left(\frac{A}{M}\right)^{(N-M+1)}\right\}P_0$$

where $P_0$ is obtained by considering all possible states of the system, i.e. $P_0, P_1 \dots P_N$

Thus

$$P_0 = 1/[1+A+A^2/2! \dots A^{(M-1)}/(M-1)! + (A^M/M!)\{M/(M-A)\} \times \{1-(A/M)^{(N-M+1)}\}].$$

Hence

$$P_D = \frac{P_L MB}{P_L MB + (M-A)(1-P_L)} \tag{3.5}$$

where $B = 1-(A/M)^{(N-M+1)}$

(b) $N = \infty$.

It is easily verified that

$$P_D = \frac{\dfrac{A^M}{M!}\left(\dfrac{M}{M-A}\right)}{1+A+A^2/2! \dots A^{(M-1)}/(M-1)! + (A^M/M!)\{M/(M-A)\}}.$$

However, if we substitute the probability of loss function $P_L$ the above reduces to

$$P_D = \frac{MP_L}{M-A(1-P_L)}. \tag{3.6}$$

### *Average Number in the Queue*

The average number in the queue is the weighted (by probabilities $P_n$) mean of the number of units queueing in state $n$. Thus for the infinite queue situation

$$\begin{aligned}\bar{L}_q &= 1 \times P_{M+1} + 2 \times P_{M+2} + \dots \\ &= \left[\frac{A^M}{M!}\left(\frac{A}{M}\right) + 2 \cdot \frac{A^M}{M!}\left(\frac{A}{M}\right)^2 + \dots\right] P_0 \\ &= \frac{A^M}{M!}\left(\frac{A}{M}\right)\frac{M^2}{(M-A)^2} \cdot P_0 \\ &= \frac{AM}{(M-A)} \cdot \frac{P_L(M-A)}{M-A(1-P_L)} \\ &= \frac{A}{M-A} \cdot P_D. \qquad (3.7)\end{aligned}$$

### *Average Delay to Units in the Queue*

It is well known in queueing theory that queue length and waiting time per unit are related in the following simple way, viz.

$$\begin{aligned}\bar{W}_q &= \bar{L}_q/\lambda \quad \text{seconds} \\ &= \bar{L}_q/q \quad \text{hours.}\end{aligned}$$

From (3.7) above

$$\begin{aligned}\bar{W}_q &= \frac{P_D \cdot A}{M-A} \cdot \frac{1}{q} \\ &= \frac{P_D \cdot \bar{T}_s}{(M-A)}. \qquad (3.8)\end{aligned}$$

The total average delay per unit $\bar{W}$ (in service and queueing) is $\bar{W}_q + \bar{T}_s$.

Although the mean waiting time is a most useful measure of performance of a traffic element, it is important also to have a measure of the probability that the wait in the queue will be greater than some arbitrary time $T$.

$$W_q(t > T) = P_D \cdot e^{-(M-A)T/\bar{T}_s}. \qquad (3.9)$$

*Results for the single-channel queueing element* ($M = 1$)

It is left to the student to verify the following results for the single-channel queue, viz.

For infinite queues

(i) When $M = 1$, $A = \rho$, $P_L = \dfrac{\rho}{1+\rho}$, $P_D = \rho$.

(ii) Probability of no units using the system

$$P_0 = (1-\rho).$$

(iii) Probability of $n$ units

$$P_n = (1-\rho)\rho^n. \tag{3.10a}$$

(iv) Average number in queue

$$\bar{L}_q = \frac{\rho^2}{1-\rho}. \tag{3.10b}$$

(v) Average number in system

$$\bar{L} = \frac{\rho}{(1-\rho)}. \tag{3.10c}$$

(vi) Average delay in queue

$$\bar{W}_q = \frac{\rho}{\mu(1-\rho)} = \frac{\rho \bar{T}_s}{1-\rho}. \tag{3.10d}$$

(vii) Average delay in system

$$\bar{W} = \frac{1}{\mu(1-\rho)} = \frac{\bar{T}_s}{(1-\rho)}. \tag{3.10e}$$

(viii) Probability of delay in queue $> T$

$$W(t > T) = \rho e^{-(1-\rho)T/\bar{T}_s}. \tag{3.10f}$$

*For queues when there is a maximum number N permitted in the system*

(i) $$P_0 = \frac{(1-\rho)}{1-\rho^{N+1}}, \qquad P_n = \frac{(1-\rho)}{1-\rho^{N+1}} \cdot \rho^n.$$

(ii) $$\bar{L}_q = \frac{1-N\rho^{N+1}+(N-1)\rho^N}{(1-\rho)(1-\rho^{N+1})} \cdot \rho^2. \tag{3.11}$$

(iii) $$\overline{W}_q = \bar{L}_q \,.\, \overline{T}_s/\rho. \tag{3.12}$$

*Non-exponential distributions of service and arrival times*

All of the above results except that for $P_L$ assume that both the service-time and the arrival-time distributions are exponential. In the case of $P_L$ the service-time distribution may have arbitrary form. A most useful formula for average delay in single queue is the Pollaczak–Khintchine formula, which takes account of an arbitrary service-time distribution.

$$\overline{W}_q = \frac{\rho}{2(1-\rho)} \overline{T}_s \left(1+\frac{\sigma_s^2}{\overline{T}_s^2}\right). \tag{3.13}$$

Remembering that the Erlang number $K = \overline{T}_s^2/\sigma_s^2$, this may be written

$$\overline{W}_q = \frac{\rho \,.\, \overline{T}_s}{2(1-\rho)} \frac{K+1}{K}. \tag{3.14}$$

When $K = \infty$, i.e. for regular service

$$\overline{W}_q = \frac{\rho \,.\, \overline{T}_s}{2(1-\rho)},$$

which is just one-half the delay for exponential service (cf. eqn. (3.10d)). However, equations for generalized arrival and service distributions are quite intractable and recourse to simulation is a most useful method of approach. An example of the simulation method is discussed later. In the meantime the results of a simulation of a $6 \times 6$ combination of Erlang service and arrival distributions is shown in Fig. 3.5. It should be noted that a $2 \times 2$ Erlang combination reduces delay as much as a

1 × 10 or 10 × 1 service/arrival combination. An Erlang 2 distribution involves very little "improvement" in either the service facility or the traffic stream.

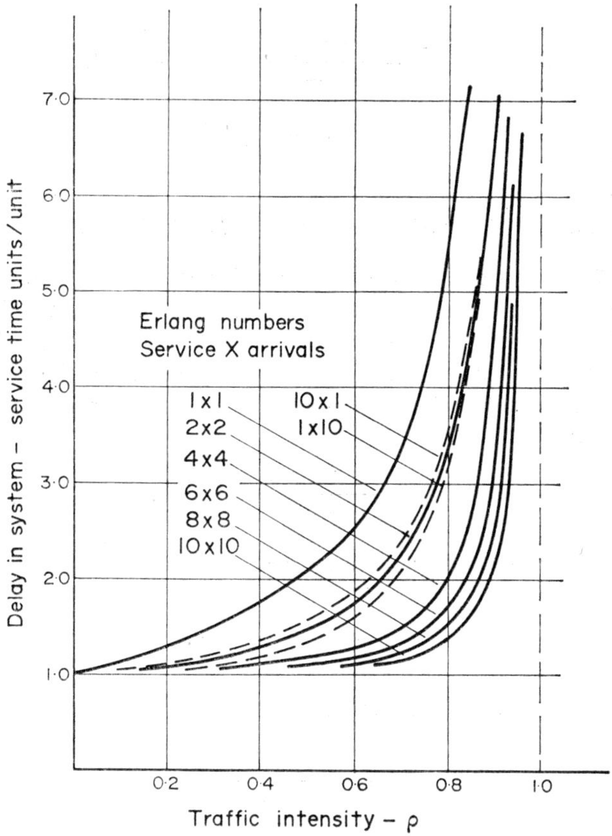


FIG. 3.5. Delay/flow characteristics of single-channel queueing system obtained by simulation.[35]

SOME GENERAL PROPERTIES OF THE QUEUEING PROCESS

Figure 3.6 shows graphically the general nature of the delay/flow function, $\overline{W}$, for a single-channel queueing element with certain "extreme" service and arrival distributions.

The important properties to note are:

(a) The delay/flow characteristic is extremely non-linear and in all cases the delay tends to infinity as the traffic intensity $\rho$ approaches unity. This produces a powerful stabilizing force on all traffic operations.

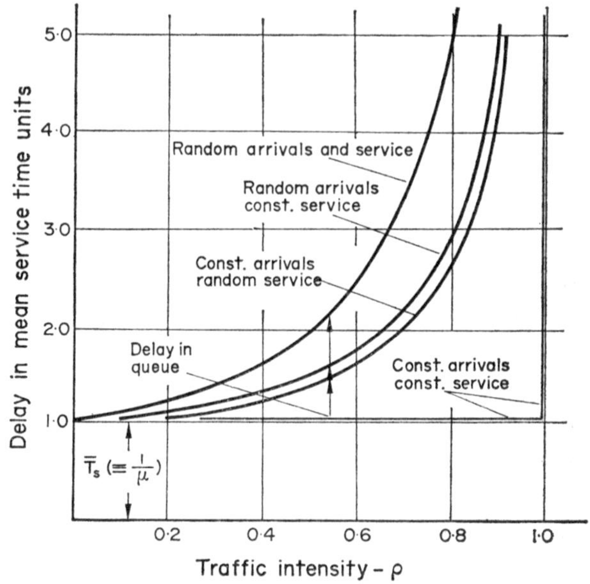


FIG. 3.6. Delay flow characteristics of single-channel traffic element for certain limiting arrival and service distributions.

(b) For low flows ($\rho \to 0$) the delay approaches the value of the average service time.

(c) Reducing the variability of the service operation or the arrival pattern reduces the delay significantly—especially at the larger values of $\rho$. In a general kind of way reduction in the variability of the service operation is synonymous with "good" design of the traffic element, e.g. bus entrances, ship holds and wharf equipment; greater uniformity in the arrival headways infers more effective traffic control, e.g. the "pushers" at Japanese railway stations!

(d) If both the service and arrival distributions are uniform there is of course no delay other than the service-time delay over the whole range from zero flow to saturation; in other words, the traffic element operates with uniform efficiency from no load to full load.

If we turn to the multi-channel queueing facility, Fig. 3.7 highlights the important general property that the efficiency of operation is enhanced quite dramatically—as the number of channels is increased.

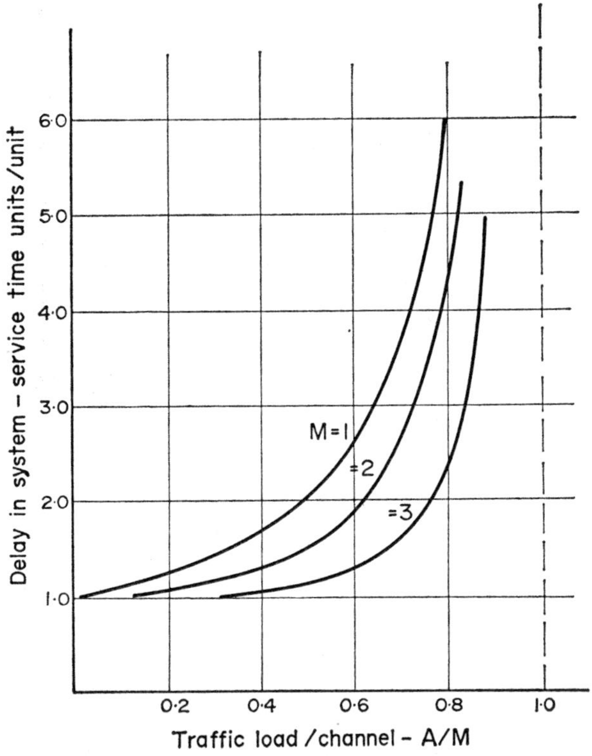


FIG. 3.7. Comparative performance of multi-channel queueing facilities with same load per channel.

***Some Illustrative Examples***

1. A transport terminal has five docks. A study of the unloading times of the delivery vans shows that they have a negative exponential distribution with a mean of 35 minutes. If they arrive at the rate of six per hour determine—

(a) the probability that an arriving van will find all docks occupied and the probability that a vehicle will be delayed;
(b) the average number of vehicles that will be queued up whilst awaiting docking space;
(c) the average time taken from arrival to departure;
(d) the capacity of reservoir space to ensure that the probability that a van will be forced to leave is not greater than 5%.

The traffic load $A = q \,.\, \overline{T}_s$

$$= \frac{6 \times 35}{60} = 3{\cdot}5.$$

Number of channels, $M = 5$.

Refer to
Table 3.1 for $(A, M) = (3{\cdot}5, 5)$ gives $P_L$ (by interpolation) $= 15{\cdot}5\%$.

The probability of delay

$$P_D = \frac{MP_L}{M - A(1 - P_L)}$$

$$= \frac{5 \times 0{\cdot}155}{5 - 3{\cdot}5(1 - 0{\cdot}155)} = 0{\cdot}33.$$

The average number of queued vans is given by

$$L_q = \frac{A}{M-A} \,.\, P_D = \frac{3{\cdot}5}{5 - 3{\cdot}5}(0{\cdot}33) = 0{\cdot}77 \text{ vans.}$$

The average time from arrival to departure is

$$W_q + T_s = L_q/q + \overline{T}_s$$

$$= 0{\cdot}77/(6/60) + 35$$

$$= 7{\cdot}7 + 35 = 42{\cdot}7 \text{ minutes.}$$

The probability that the queue will be as long or longer than $N$ vans is

$$
\begin{aligned}
P(\text{queues} \geqslant N) &= \sum_{n=0}^{\infty} P_{M+N+n} \\
&= \left[\frac{A^M}{M!}\left(\frac{A}{M}\right)^N + \frac{A^M}{M!}\left(\frac{A}{M}\right)^{N+1} + \ldots\right] P_0 \\
&= \frac{A^M}{M!} \cdot \left(\frac{A}{M}\right)^N . P_0 \left[1 + \frac{A}{M} + \left(\frac{A}{M}\right)^2 + \ldots\right] \\
&= \frac{A^M}{M!}\left(\frac{A}{M}\right)^N \frac{M}{M-A} P_0 \\
&= \left(\frac{A}{M}\right)^N P_D \\
&= (0{\cdot}7)^N (0{\cdot}33).
\end{aligned}
$$

Therefore

$$
\begin{aligned}
P(\text{queue} \geqslant 1) &= 0.23 \\
&\ldots\ldots\ldots \\
P(\text{queue} \geqslant 5) &= 0{\cdot}055 \\
P(\text{queue} \geqslant 6) &= 0{\cdot}018.
\end{aligned}
$$

It would thus be necessary to have six reservoir spaces, but five would be a fair practical answer.

2. Table 3.2 summarizes the results of 220 observation times at a kerbside bus stop. What is the maximum rate of flow of buses through the stop? Plot the delay/flow characteristic of the bus stop.

The maximum rate of flow of buses through the stop is therefore

$$q_{\max} = \frac{3600}{\overline{T}_s} = \frac{3600}{29{\cdot}7} = 121 \text{ buses/hour.}$$

To obtain the delay flow characteristic we compute Table 3.3 using the Pollaczak–Khintchine formula, i.e.

$$\overline{W}_q = \frac{\rho}{2(1-\rho)} \cdot \overline{T}_s \left(1 + \frac{\sigma_s^2}{T_s^2}\right).$$

Here

$$\bar{T}_s = 0{\cdot}495 \text{ minutes}$$

and

$$\sigma^2/\bar{T}^2 = 267/(29{\cdot}7)^2 = 267/882 = 0{\cdot}303.$$

TABLE 3.2. FREQUENCY DISTRIBUTION OF LOADING/UNLOADING TIMES FOR A SAMPLE OF BUSES AT A CENTRAL SYDNEY BUS STOP

| Loading/unloading times—$t$ (sec) | frequency $f$ | $f \times t$ | $f \times t^2$ |
|---|---|---|---|
| 0–5·9 | 4 | 12 | 36 |
| 6–11·9 | 19 | 171 | 1539 |
| 12–17·9 | 32 | 480 | 7200 |
| 18–23·9 | 44 | 924 | 19,404 |
| 24–29·9 | 33 | 891 | 24,057 |
| 30–35·9 | 22 | 726 | 23,958 |
| 36–41·9 | 17 | 663 | 25,857 |
| 42–47·9 | 16 | 720 | 34,200 |
| 48–53·9 | 11 | 561 | 28,611 |
| 54–59·9 | 9 | 513 | 29,241 |
| 60–65·9 | 6 | 378 | 23,814 |
| 66–71·9 | 5 | 345 | 23,805 |
| 72–77·9 | 2 | 150 | 11,250 |
| 78 | 0 | 0 | 0 |
| | 220 | 6534 | 252,972 |

Therefore $\bar{T}_s = 29{\cdot}7$ sec and $\sigma_s^2 = 267$ sec$^2$.

TABLE 3.3.

| Flow $q$/hr | Traffic load $A\ (=\rho)$ | $\bar{W}_q$ (min) | $\bar{W}$ (min) |
|---|---|---|---|
| 0 | 0 | 0 | 0·50 |
| 20 | 0·17 | 0·07 | 0·57 |
| 40 | 0·33 | 0·12 | 0·62 |
| 60 | 0·50 | 0·32 | 0·82 |
| 80 | 0·66 | 0·64 | 1·14 |
| 100 | 0·83 | 1·57 | 2·07 |

These results may be plotted to give the delay/flow characteristic shown in Fig. 3.4(e).

The Erlang number is 3 (i.e. integral part of 1/0·303). The student should develop the cumulative distribution of Table 3.2, plot it and compare it with an Erlang curve with $K = 3$. The student might also form the frequency function of 3.1 and evaluate it for the fractional value of $\overline{T}_s^2/\sigma_s^2$.

## *Some Special Queue Problems*

### *Queueing for gaps*

The problem of "access" to frontage land uses (e.g. service stations, supermarkets, motels, even whole subdivisions) from a main or distributor road is of great relevance in land-use and transport planning.

Mathematically the problem is one of queueing for gaps in a priority traffic stream—exiting vehicles are in general required to give way to main-road traffic in both directions; the right-turn entries must give way to the opposing traffic stream; the left-turn entries must give way to pedestrians on the footpath.

### *Acceptance gaps*

In this context an acceptance gap is the headway required by a vehicle (at rest or moving slowly) to merge into or cross a free moving stream of traffic. A variety of research on this topic[20–23] has established the general character of the acceptance gap distribution for urban traffic. It has a gamma or Erlang form with a mean of 4–5 seconds and a variance of approximately 2 seconds. The acceptance gap is usually associated with an isolated vehicle seeking to enter or cross a traffic stream. When a number of vehicles are queued waiting for gaps then in cases where the main stream headways are large enough for a number of vehicles to enter the follow-up headways for the second, third, etc., vehicle are of the order of 2–3 seconds.

There is an extensive literature of a rigorous mathematical kind pertaining to this problem[24–26] and the work of Tanner is notable. The following results are sufficient for most practical operational and planning problems.

*The absorption rate of a priority traffic stream*

Suppose that a long queue is waiting to enter a main traffic stream, e.g. a parking station waiting to discharge. Then if the flow in the traffic lane to be entered is $q$/hour, the absorption rate $p$ is

$$p = q\frac{e^{-qT/3600}}{1-e^{-qt/3600}} \tag{3.15}$$

where $T$ is the acceptance gap for the first vehicle to enter an acceptable headway and $t$ is the follow-up headway. Table 3.4 gives values of $p$ for a limited range of the parameters $T$ and $t$.

TABLE 3.4. ABSORPTION RATE OF A PRIORITY TRAFFIC STREAM

| Main stream $q$/hr | Absorption rate $p$ (veh./hr) | | | |
|---|---|---|---|---|
| | $T = 4$ sec | | $T = 5$ sec | |
| | $t = 2$ | $t = 3$ | $t = 2$ | $t = 3$ |
| 200 | 1522 | 1043 | 1440 | 986 |
| 400 | 1289 | 904 | 1151 | 809 |
| 600 | 1086 | 782 | 919 | 662 |
| 800 | 916 | 675 | 733 | 541 |
| 1000 | 772 | 582 | 584 | 441 |
| 1200 | 650 | 500 | 465 | 358 |
| 1400 | 546 | 429 | 370 | 290 |
| 1600 | 459 | 367 | 294 | 235 |
| 1800 | 385 | 313 | 233 | 190 |

*Delay to an isolated vehicle entering or crossing a priority stream*

The average waiting time, $\overline{W}_q$, for isolated vehicles is

$$\overline{W}_q = \frac{3600}{q}\{e^{qT/3600}-1\}-T. \tag{3.16}$$

This is generally known as Adams formula. He presented it in his celebrated paper on the stochastic character of the traffic stream published in 1936. Later Tanner developed the statistical distribution of which the above formula is the mean. From the distribution the variance has been evaluated.

$$\text{Var. }(W_q) = \left(\frac{3600}{q}\right)^2 \left\{e^{qT/1800} - \frac{qT}{1800}e^{qT/3600} - 1\right\}. \tag{3.17}$$

The distribution itself is of considerable usefulness in dealing with access problems because it is often necessary to know the probability of occurrence of the longer waiting times, especially if the delayed vehicle is holding up main stream traffic (i.e. the right-turn entry case). Table 3.5 summarizes the distribution and provides probabilities of waiting longer than an arbitrarily chosen interval $W$.

*Probability of delay in gaining access to the traffic stream*

$P_D$, the probability of being delayed in seeking entry to the traffic stream is

$$1 - P(\text{gap} > T), \quad \text{i.e.} \quad P_D = 1 - e^{-qT/3600}. \tag{3.18}$$

Tanner has produced further results which give the delay suffered by access vehicles when the access rate is such as to produce significant queueing effects, i.e. when greater than about 200 vehicles/hour. The analytical results are algebraically cumbersome but for practical purposes their graphical representation, Figs. 3.8(a) and (b) provide adequately for the engineer and planner.

It should be noted that all of the results in this section are based on the assumption that the priority traffic stream has a negative exponential distribution of headways.

### *An Example*

A small traffic generator draws customers from the passing traffic on a busy two-lane, two-way road. The peak traffic flow is 800 vehicles/hour in the lane adjacent to the development and 400 vehicles/hour in the opposite direction. The vehicular traffic intensity generated by the development is expected to be about 90 vehicles/hour in the hour coinciding with the peak flow on the frontage road.

TABLE 3.5. PROBABILITY OF AN ISOLATED VEHICLE BEING DELAYED GREATER THAN $W$ SECONDS

(Note: if $q$ is measured in units/hr the $T$ must be expressed in hours)

| $W/T$ \ $q.T$ | 0·25 | 0·50 | 0·75 | 1·0 | 1·5 | 2·0 | 3·0 | 4·0 | 5·0 |
|---|---|---|---|---|---|---|---|---|---|
| 0 | 0·221 | 0·394 | 0·528 | 0·632 | 0·777 | 0·865 | 0·950 | 0·982 | 0·993 |
| 0·5 | 0·124 | 0·242 | 0·350 | 0·448 | 0·610 | 0·729 | 0·876 | 0·945 | 0·976 |
| 1·0 | 0·026 | 0·070 | 0·173 | 0·264 | 0·442 | 0·594 | 0·801 | 0·908 | 0·960 |
| 2·0 | | 0·017 | 0·049 | 0·099 | 0·238 | 0·397 | 0·670 | 0·839 | 0·927 |
| 4·0 | | | 0·004 | 0·014 | 0·068 | 0·176 | 0·469 | 0·716 | 0·864 |
| 6·0 | | | | 0·004 | 0·020 | 0·079 | 0·320 | 0·625 | 0·813 |
| 8·0 | | | | | 0·006 | 0·035 | 0·228 | 0·538 | 0·760 |
| 10·0 | | | | | | 0·016 | 0·159 | 0·462 | 0·711 |
| 15·0 | | | | | | | 0·065 | 0·317 | 0·602 |

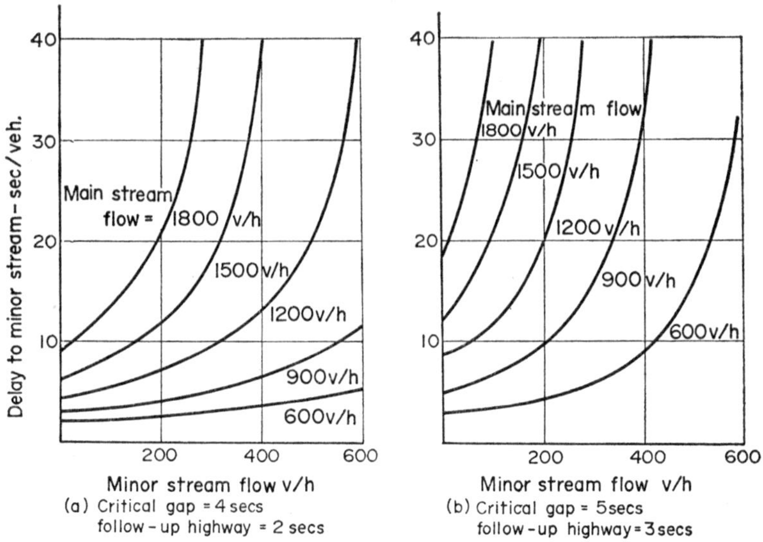


FIG. 3.8. Delay to vehicles wishing to enter or cross a priority traffic stream.[19]

Determine

(a) the probability that a departing vehicle will be delayed;
(b) the average delay to all departing vehicles;
(c) the average delay to entering vehicles;
(d) the probability that a right-turn entry* will be delayed longer than 10 seconds;
(e) the effect on the capacity of the frontage road;
(f) the absorption capacity of the frontage road for left-turning and right-turning departures.

Figure 3.9 summarizes the traffic situation on the assumption that the development draws traffic from the passing stream in proportion to the directional flows.

### *Probabilities and Delays to Departing Vehicles*

If left-turn departures require a gap 4 seconds or greater in a flow of 800 vehicles/hour, then from (3.18)

$$P_D = 1-e^{-qT/3600} = 1-e^{-(800\times 4)/3600}$$
$$= 1-0{\cdot}41 = 0{\cdot}59.$$

If we assume right-turn departures require a slightly longer gap—5 seconds, say—this must be taken in the combined traffic stream, i.e. $q = 1200$ vehicles/hour and

$$P_D = 1-e^{-(1200\times 5)/3600}$$
$$= 1-0{\cdot}19 = 0{\cdot}81.$$

The delay suffered by departing vehicles may be determined by using 3.16, i.e.

$$\overline{W}_L = \frac{3600}{q}e^{+qT/3600}-\frac{3600}{q}-T.$$

For left turners

$$T = 4 \quad \text{and} \quad q = 800.$$

* Note: directions of turns should be interpreted with respect to the left-side driving rule.

Therefore,

$$\overline{W}_R = 4{\cdot}5 \times \frac{1}{0{\cdot}41} - 4{\cdot}5 - 4 = 2{\cdot}5 \text{ sec.}$$

For right turners

$$T = 5 \quad \text{and} \quad q = 1200.$$

Therefore,

$$\overline{W} = 3 \times \frac{1}{0{\cdot}19} - 3 - 5 = 7{\cdot}9 \text{ sec.}$$

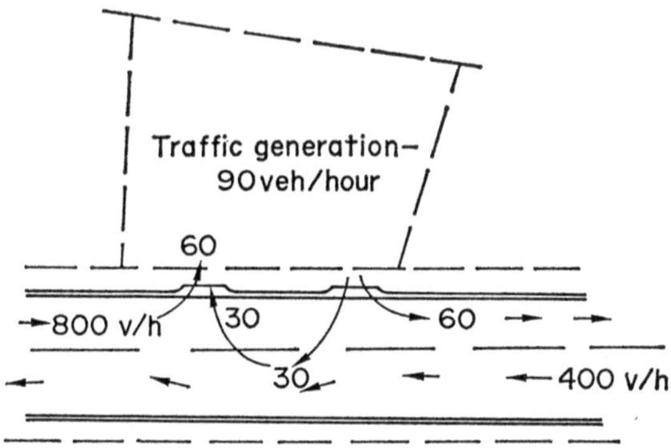


FIG. 3.9. Diagram of entry/exit movements for small traffic generator drawing traffic from passing traffic stream (frontage road is undivided and access movements are proportional to directional flows).

The average delay may be taken as the weighted average, i.e.

$$\tfrac{2}{3} \,.\, \overline{W}_L + \tfrac{1}{3}\overline{W}_R = \tfrac{2}{3} \,.\, 2{\cdot}5 + \tfrac{1}{3} \,.\, 7{\cdot}9 = 4{\cdot}5 \text{ sec.}$$

If queueing effects are taken into account then with resort to Fig. 3.7 (a) and (b)

$$\overline{W}_L = 2{\cdot}8 \text{ sec} \quad \text{and} \quad \overline{W}_R = 9{\cdot}5 \text{ sec.}$$

### *Delay to Entering Vehicles*

Provided there is no queueing resulting from overloading of the parking facilities left-turn entries do not in general cause any delay to main-stream vehicles. On the other hand, right-turn entries in waiting for a gap in the opposing traffic interact directly with other traffic in their lane.

In this example the right-turn entries need to find a gap in the opposing flow of 800 vehicles/hour. If the critical gap is taken as 4 seconds then the calculations are similar to those already made for the left-turn exits, i.e. the average delay to entering vehicles is 2·5 seconds.

The probability that an entering vehicle will be delayed 10 seconds or more can be evaluated from Table 3.4. $W/T = 10/4 = 2{\cdot}5$ and $qT/3600 = 0{\cdot}9$ (approx.). Therefore

$$P(W > 10) = 0{\cdot}058, \text{ i.e. about } 6\%.$$

### *The Effect of Right Turners on Capacity*

If the number of right-turn entries is thirty and their average delay is 2·5 seconds then the total time they spend "blocking" the lane is $30 \times 2{\cdot}5 = 75$ seconds. The effective capacity of the lane is thus

$$\frac{3600-75}{3600} \times 100 \times \text{original capacity} = 98\% \text{ original capacity.}$$

### *Absorption Capacity*

Reference to Table 3.3 gives for

left-turn exits with $T = 4, t = 2, p = 916$ vehicles/hour,
right-turn exits with $T = 5, t = 3, p = 358$ vehicles/hour.

## TRAVEL-TIME/FLOW RELATIONSHIPS FOR ROAD TRAFFIC NETWORK ELEMENTS

The main elements of a road traffic network are:

(a) uninterrupted length of roadway—usually marked into lanes, i.e. motorways, parkways, some major at-grade arterials;

(b) lengths of at-grade roadways subject to various impediments, e.g. pedestrian crossings, minor intersections, access from frontage land uses, parking;

(c) bottleneck situations—three lanes merging into two, reduced geometric standards of individual lanes;
(d) major traffic-light-controlled intersections;
(e) on ramps and merging lanes;
(f) pedestrian crossing facilities.

The bottleneck situation is comparatively easily dealt with by considering a barrier to exist at the commencement of the bottleneck section and associating with it a service rate equal to the saturation flow of the bottleneck section. Single-channel queueing results can then be usefully applied. It is instructive to apply shock-wave theory of fluid flow in some circumstances.[27]

A simple analysis of on-ramp and merging problems can be made by using the absorption rate formula (3.13) to obtain a "service rate" and then applying the simple single-channel queueing formulae.

## INTERSECTIONS

There is now an extensive literature on many aspects of intersection behaviour but even so the control of traffic at intersections remains as yet in a relatively undeveloped state.

It is not intended to exploit this complex subject in this work. However, it is appropriate to provide the student and analyst with guidance as to how to determine the contribution made by traffic-signal-controlled intersections to the total travel time on any path linking an origin and a destination, given the operational characteristics of the traffic control system employed.

## DELAY IN AN INTERSECTION APPROACH WHEN ARRIVALS ARE RANDOMLY SPREAD OVER THE WHOLE CYCLE

An exact solution to this problem has not yet been obtained. The service time is in fact time-dependent and the transient solutions required are not amenable to algebraic manipulation. However, a number of good approximations have been obtained—first by Webster, then by Miller, Newell, Blunden and others.[28–32]

Webster's approximation is

$$W = 0{\cdot}9\left[\frac{(1-a)^2C}{2(1-ay)}+\frac{3600y^2}{2q(1-y)}\right]. \tag{3.19}$$

The factor 0·9 is an empirically-determined correction factor.

Blunden—

$$W = \frac{(1-a)^2C}{2}+\frac{3600y^2}{2q(1-y)}\cdot\left(\frac{K+1}{K}\right). \tag{3.20}$$

Miller—

$$W = \frac{(1-a)^2C}{2(1-ay)}+\frac{(1-a)(2y-1)}{2(1-ay)(1-y)}+\frac{a^3y^2}{2q(1-ay)}, \tag{3.21}$$

where $C$ = cycle time,

$a$ = the effective green time ratio,

$y$ = degree of saturation = $q/as$,

$K$ = the Erlang number of the start-up headways—ranges from six with 25% commercial vehicles to nine for cars only,

$s$ = saturation flow or starting-up rate of a line of vehicles—about 1800 per hour for cars only to 1200 per hour for 25% C.V.'s,

$q$ = the average arrival flow—vehicles/hour,

and $\overline{W}$ = the average delay per vehicle—seconds.

Consider a set of signals with a cycle time of 60 seconds and an effective green time of 30 seconds. Therefore $a = 0{\cdot}5$. If the distribution of start-up headways has a mean of 2 seconds and if the $K$ number of their distribution is 9, determine the delay suffered by vehicles in a single lane arriving at the rate of 200, 400, 600 and 800 per hour. The results are tabulated in Tables 3.6–3.8.

TABLE 3.6. WEBSTER'S FORMULA[28]

| Approach flow Veh./hr | $y$ | 1 | 2 | $W$ |
|---|---|---|---|---|
| | | $\frac{(1-a)^2C}{2(1-ay)}$ | $\frac{3600y^2}{2q(1-y)}$ | $0{\cdot}9(1+2)$ |
| 200 | 0·22 | 8·5 | 0·6 | 8·2 |
| 400 | 0·45 | 9·6 | 1·6 | 10·1 |
| 600 | 0·67 | 11·3 | 3·9 | 13·7 |
| 800 | 0·90 | 13·5 | 16·0 | 26·5 |

TABLE 3.7. BLUNDEN'S MODIFICATION—$K = 9$[31]

| Approach flow Veh./hr | $y$ | 1 | 2 | $W$ |
|---|---|---|---|---|
| | | $\frac{(1-a)^2C}{2}$ | $\frac{3600y^2}{q(1-y)} \cdot \frac{K+1}{K}$ | $(1+2)$ |
| 200 | 0·22 | 7·5 | 0·6 | 8·1 |
| 400 | 0·45 | 7·5 | 1·78 | 9·3 |
| 600 | 0·67 | 7·5 | 4·35 | 11·9 |
| 800 | 0·90 | 7·5 | 17·8 | 25·3 |

TABLE 3.8. MILLER'S FORMULA[29]

| Approach flow Veh./hr | $y$ | 1 | 2 | 3 | $W$ |
|---|---|---|---|---|---|
| | | $\frac{(1-a)^2C}{2(1-ay)}$ | $\frac{(1-a)(2y-1)}{2(1-ay)(1-y)}$ | $\frac{a_3y^2}{2(1-ay)}$ | $(1+2+3)$ |
| 200 | 0·22 | 8·5 | 0 | — | 8·5 |
| 400 | 0·45 | 9·6 | 0 | 0·1 | 9·7 |
| 600 | 0·67 | 11·3 | 4·5 | 0·25 | 16·1 |
| 800 | 0·90 | 13·5 | 16·0 | 0·40 | 29·9 |

The results of Tables 3.6–3.8 are shown plotted in Fig. 3.10.

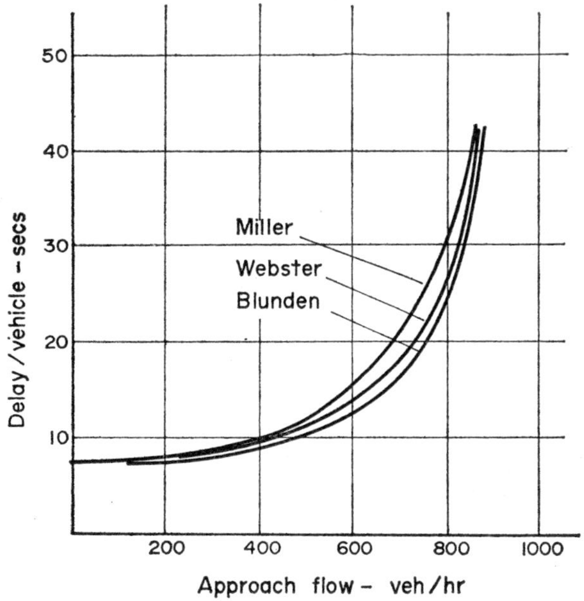


FIG. 3.10. Delay flow results for an intersection approach as determined by Webster, Miller, Blunden formulae.

## THE TRAVEL-TIME/FLOW RELATIONSHIP FOR FINITE LENGTHS OF ROAD

As yet there is no complete theoretical explanation of the relationship between travel time and flow for a finite length of road. There is, of course, a relationship between the saturation flow and speed and it has been a common practice to regard the reciprocal of the speed variable as being proportional to the travel time per unit distance. This is not true. The delay in this context arises from the disturbance to the free speed distribution resulting from faster vehicles catching up on slower ones and being required to wait until passing or lane-changing is possible or, in the case of single-lane or crowded lane conditions, being prevented from passing at all. On this basis there is undoubtedly a queueing mechanism at work. Certainly all the experimental evidence points to the validity of the familiar queueing type curves for delay and flow. Observed flow situations are shown in Figs. 3.11 (a), (b) and (c). A number of well-defined characteristics of these experimental plots may be stated, viz.

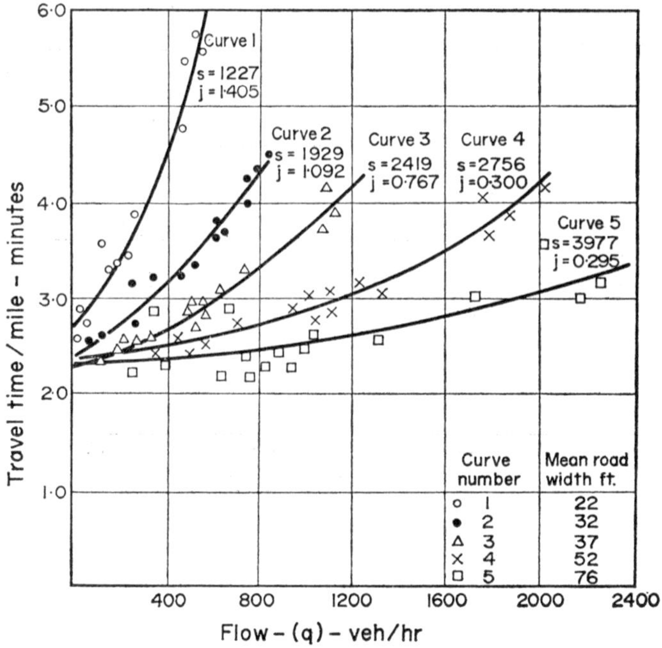


FIG. 3.11(a). Travel-time/flow relationship fitted to data collected by Wardrop for different types of streets in central London.(36,37)

(a) At "zero" flow the intercept on the travel-time axis is well defined and in the case of road traffic represents the reciprocal of the mean of the free speed distribution. It is a characteristic of the "scheduled" working of all transport systems and is synonymous with the mean service time in queueing theory.

(b) In the steady state the delay is asymptotic to the saturation flow ordinate.

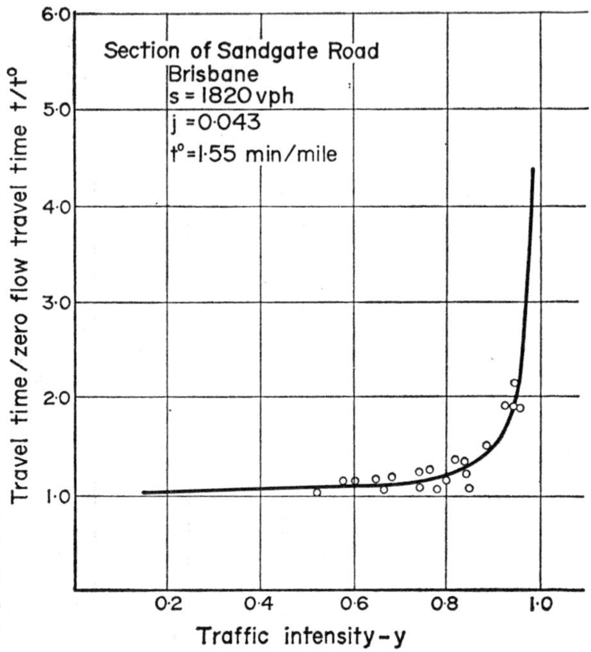


FIG. 3.11(b). Travel-time/flow relationship fitted to data obtained on a section of Sandgate Road, Brisbane.(36)

(c) The curve itself is monotonically increasing and the more "smooth" the flow situation the smaller the resulting travel time at least until saturation is imminent.

A family of curves with these three properties may be represented by

$$A+B\frac{f(y)}{1-y}.$$

Translating this into a notation appropriate to flow phenomena we may write the travel time, $t$, per unit distance as

$$t = t^0\,\frac{1-(1-j)y}{1-y} \tag{3.22}$$

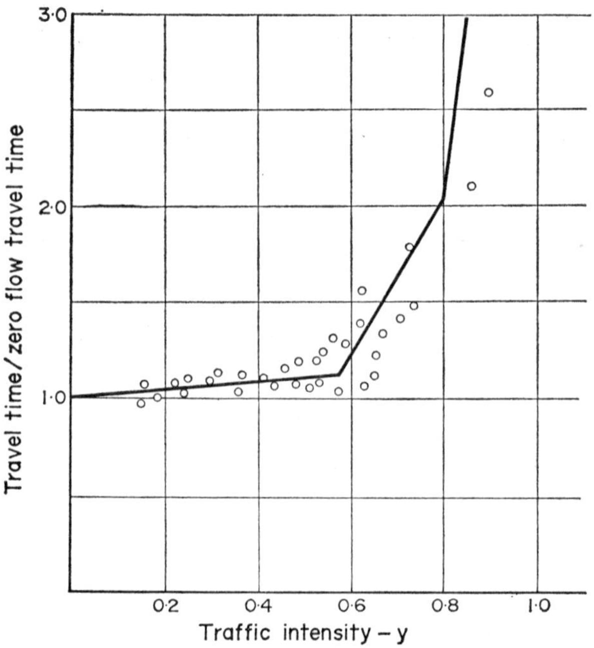


FIG. 3.11(c). Travel-time/flow data and curve used in Toronto transportation study.

where $t^0$ = zero flow travel time,

$y$ = $q/s$ the ratio of the flow to the saturation flow,

$j$ = level of service factor and is a function of the factors causing variability in the flow situation, i.e. parked vehicles, minor intersection, pedestrian crossings and so on.

Comparing this with the general formula given

$$A = t^0,$$
$$B = j \times t^0,$$
$$f(y) = y,$$

when $j = 1$ the expression for $t$ corresponds with the single-channel queueing formula for exponential arrivals and service. When $j = 0$, then $t = t^0$ and there is no queueing delay at all.

In the absence of authentic data it seems appropriate in making initial evaluations or theoretical assessments to assume the values shown in Table 3.9 for the parameters of the travel-time/flow equation.

TABLE 3.9. FLOW PARAMETERS FOR VARIOUS ROAD TYPES

| Conditions | $t^0$ (min/mile) | "$j$"-value | Saturation flow (veh./hr) |
|---|---|---|---|
| Motorways | 0·8–1·0 | 0–0·2 | 2000/lane |
| Multi-lane urban arterials | 1·5–2·0 | 0·4–0·6 | 1800/lane |
| Feeder and collector roads | 2·0–3·0 | 1·0–1·5 | 1800/total width |

## SIMULATION OF QUEUEING PHENOMENA

As indicated earlier, the queueing situation is easy to simulate on a digital computer. There are several reasons why simulation is particularly useful in this situation:

(a) As queueing phenomena are essentially discrete flow problems they can be simulated in a more or less literal sense on a digital computer.

(b) Although the results we have obtained in what has gone before are elegant in the extreme, they have been restricted in the main to situations where the arrivals have been exponential and the service times either exponential or constant, except in the case of the single-channel system where a more general distribution was treated. The treatment of general distributions of arrival service times leads to equations which are usually intractable. Simulation programmes once worked out can accept any form of input and output;

(c) it is a powerful research tool.

### *The Method of Simulating a Queueing System*

Queueing phenomena are easy to simulate. The integral transform theorem of probability theory provides the basis for the method. Consider a service time (or an arrival interval) having a continuous distribution.

If $s(t)$ is its density function and $S(t)$ the cumulative distribution, then the transformation

$$y = S(t) = \int_{t}^{\infty} s(t)\,dt \quad \text{or} \quad t = S^{-1}(y) \tag{3.23}$$

converts $t$ into a new variate, $y$, with a one to one correspondence with $t$ in the interval 0 to 1.

If now $y$ is considered to arise from a uniform distribution—which is essentially a set of random numbers—then for every random number selected in a series of trials a random value of a variate with any density function can be obtained by finding the inverse of its cumulative distribution.

This undoubtedly obscure argument can best be clarified with an example. We will start by considering the set of loading/unloading times already given in Table 3.2.

The frequency histogram can be constructed and cumulative distribution of the loading/unloading times may be approximated by a smooth curve—Fig. 3.12. It is appropriate to truncate the distribution at a convenient maximum value—in this case 72 seconds. Now if a set of random numbers between 0 and 99 are written down (from a table of random numbers or the last two digits of a set of consecutive telephone numbers from the telephone directory) as values of the ordinate, the corresponding values on the horizontal axis represent a series of randomly chosen service times. For example, suppose the first random number was 63, then the corresponding service time is 21 seconds. If the next had been 27 then the service time would be 38·5 seconds.

In some cases it is possible to determine a sample of service or arrival times analytically. This is particularly easy in the case of an exponential distribution and this is fortunate because arrivals in so many queueing problems are, for all practical purposes, random.

Let $N$ be the per cent of arrival headways between successive buses travelling along a city street that are greater than $t$ seconds. Then

$$N = 100e^{-qt/3600}. \tag{3.24}$$

Suppose $q = 83$ buses/hour, then

$$\frac{N}{100} = e^{-t/43\cdot5} \quad \text{or} \quad \log_e \frac{N}{100} = \frac{-t}{43\cdot5}$$

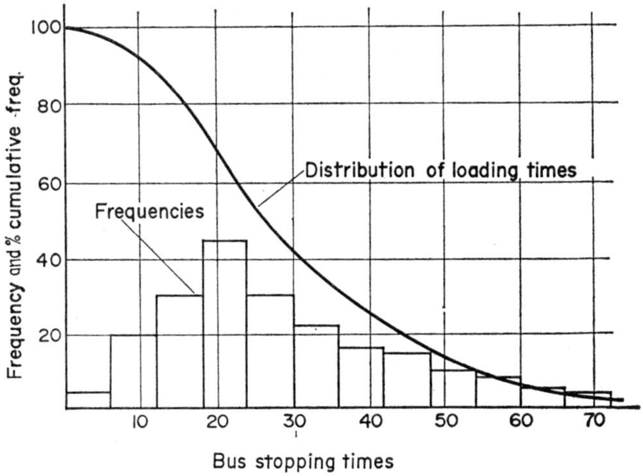


FIG. 3.12. Frequencies and distribution of kerbside bus-loading times, Pitt Street, Sydney.

or

$$t = 43\cdot5 \log_e \frac{100}{N}$$

$$= 2\cdot30 \times 43\cdot5 \log_{10} (100/N)$$

$$= 100 (\log 100 - \log N)$$

$$= 100 (2 - \log N).$$

Once again $N$ is synonymous with the random number series.

Suppose $N$ is 49: then $\log N = 1\cdot69$ and $t = 31$ seconds.

## A Manual Simulation Example

With the service and arrival distributions discussed above we can proceed to study the queueing situation that arises at a bus stop with bus loading and unloading times approximated by the distribution given in Fig. 3.12 and a flow of buses through the stop of 83/hour.

Table 3.10. Generation of Arrival and Service Times

| Arrivals | | | | | | Service times | | |
|---|---|---|---|---|---|---|---|---|
| Ser. | $N_a$ | log $N$ | 2−log $N$ | Headways $t_a$ | Arrival times $t_a$ | Ser. | $N_s$ | $t_s$ |
| 1 | | | | | 0 | 1 | 81 | 16 |
| 2 | 32 | 1·51 | 0·49 | 49 | 49 | 2 | 74 | 18 |
| 3 | 99 | 1·99 | 0·01 | 1 | 50 | 3 | 53 | 25 |
| 4 | 61 | 1·79 | 0·21 | 21 | 71 | 4 | 34 | 34 |
| 5 | 16 | 1·20 | 0·80 | 80 | 151 | 5 | 45 | 28 |
| 6 | 66 | 1·82 | 0·18 | 18 | 169 | 6 | 50 | 26 |
| 7 | 10 | 1·00 | 1·00 | 100 | 269 | 7 | 44 | 29 |
| 8 | 49 | 1·69 | 0·31 | 31 | 300 | 8 | 14 | 50 |
| 9 | 83 | 1·92 | 0·08 | 8 | 308 | 9 | 55 | 24 |
| 10 | 12 | 1·08 | 0·92 | 92 | 400 | 10 | 36 | 32 |
| 11 | 36 | 1·56 | 0·44 | 44 | 444 | 11 | 10 | 54 |
| 12 | 31 | 1·49 | 0·51 | 51 | 495 | 12 | 63 | 22 |
| 13 | 92 | 1·96 | 0·04 | 4 | 499 | 13 | 97 | 7 |
| 14 | 68 | 0·90 | 1·10 | 110 | 609 | 14 | 42 | 29 |
| 15 | 74 | 1·87 | 0·13 | 13 | 622 | 15 | 92 | 10 |
| 16 | 84 | 1·92 | 0·08 | 8 | 630 | 16 | 36 | 23 |
| 17 | 33 | 1·52 | 0·48 | 48 | 678 | 17 | 50 | 26 |
| 18 | 16 | 1·20 | 0·80 | 80 | 758 | 18 | 05 | 72 |
| 19 | 28 | 1·45 | 0·55 | 55 | 813 | 19 | 41 | 29 |
| 20 | 64 | 1·81 | 0·19 | 19 | 832 | 20 | 14 | 49 |

Table 3.10 provides a sample of twenty random arrival and service times generated from the exponential distribution and the experimental distribution shown in Fig. 3.12, respectively, from a simple set of twenty two-digit random numbers (0–99).

It will be noted that a cumulative sum of the arrival times is included and this may be plotted as shown in curve 1 of Fig. 3.13. The horizontal scale is clock time (TA) and each step in the vertical scale represents the arrival of a bus at the stop. However, it is not possible to accumulate the sequence of service times in the same manner for the simple reason that whenever the queue runs out one must break the time scale and restart it again when the next unit arrives. Curve 2 gives a plot of the departure times with each successive departure being marked by a step in the vertical scale.

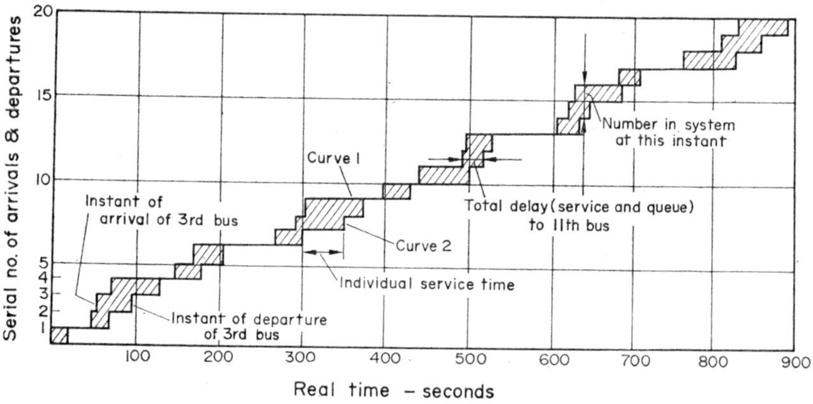


FIG. 3.13. Graphical representation of the simulation process.

When these two curves are constructed the properties of the queueing system may be ascertained as follows:

(a) the vertical interval between the two curves represents the number of the system at any given time point on the clock scale;
(b) the horizontal separation of the curves gives the delay suffered by any given unit;
(c) the area between the curves gives the total delay in the system and when divided by the total number in the sample the delay per unit (queue + service) is obtained.

The arithmetic involved can be systematized as shown in the flow diagram in Fig. 3.13. This diagram provides the basis for a computer programme.

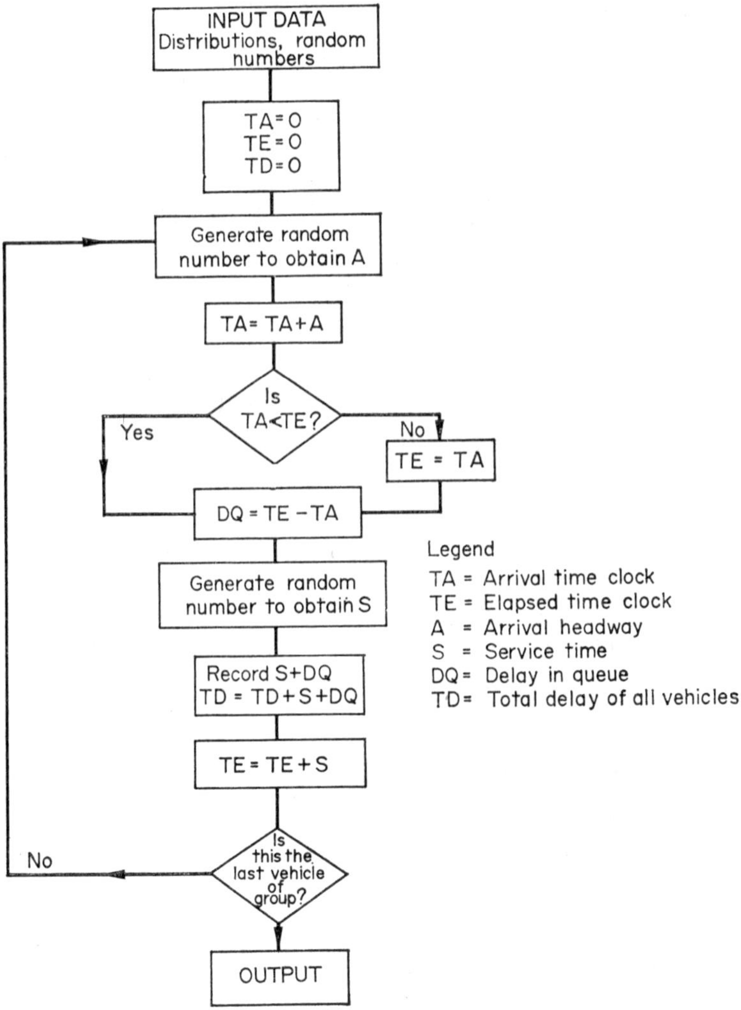


FIG. 3.14. Flow diagram for simulation of a single-channel queue system.

We may use it to evaluate a few steps in the simulation of the situation given above. With the symbols as given in the legend in Fig. 3.14 we may proceed thus:

**Initially** $TA_0 = 0 \quad TE_0 = 0 \quad TD_0 = 0.$

**Step 1**

$A_1 = 0$ (i.e. first vehicle arrives at time 0),

$TA_1 = TA_0 + A_1 = 0+0 = 0,$

$TE_0 = 0$ and $TA_1 \nless TE$ so put $TE = TA = 0,$
↑— (note initial conditions)

$DQ = TE - TA = 0-0 = 0,$

$S_1 = 16$ (from Table 2),

$S_1 + DQ = 16,$

$TD_1 = TD_0 + S_1 + DQ = 16,$
↑— (= 0, from initial conditions)

$TE_1 = TE_0 + S_1 = 16.$

**Step 2**

$A_2 = 49,$

$TA_2 = TA_1 + A_2 = 0+49 = 49,$

$TE_1 = 16$ and $TA_2 \nless TE_1$ so put $TE_1 = 49,$

$DQ = TE_1 - TA_2 = 49-49 = 0,$

$S_2 = 18$ (second generated service time),

$TD_2 = TD_1 + S_2 + DQ = 16+18 = 34,$

$TE_2 = TE_1 + S_2 = 49+18 = 67.$

**Step 3**

$A_3 = 1,$

$TA_3 = TA_2 + A_3 = 49+1 = 50,$

$TE_2 = 67$ and $TA_3 < TE_2$, so go straight to next step,

$DQ = TE_2 - TA_3 = 67-50 = 17,$

$S_3 = 25,$

$S_3 + DQ = 25+17 = 42,$

$TD_3 = TD_2 + S_3 + DQ = 34+42 = 76,$

$TE_3 = TE_2 + S_3 = 67+25 = 92.$

**Step 4** $A_1 = 21,$

$TA_4 = TA_3 + A_1 = 50 + 21 = 71,$

$TE_3 = 92$ and $TA_1 < TE_3$, so go straight to next step,

$DQ = TE_3 - TA_1 = 92 - 71 = 21,$

$S_4 = 34,$

$S_4 + DQ = 34 + 21 = 55,$

$TD_4 = TD_3 + S_4 + DQ = 76 + 55 = 131,$

$TE_4 = TE_3 + S_4 = 92 + 34 = 126.$

**Step 5** $A_5 = 80,$

$TA_5 = TA_4 + A_5 = 71 + 80 = 151,$

$TE_4 = 126$ and $TA_5 \nless TE_4$, so put $TE_4 = 151,$

$DQ = 151 - 151 = 0,$

$S_5 = 28,$

$S_5 + DQ = 28,$

$TD_5 = TD_4 + S_5 + DQ = 131 + 28 = 159,$

$TE_5 = TE_4 + S_5 = 151 + 28 = 179.$

**Step 6** $A_6 = 18,$

$TA_6 = TA_5 + A_6 = 151 + 18 = 169,$

$TE_5 = 179$, $TA_6 < TE_5$, and go straight to next step,

$DQ = TE_5 - TA_6 = 179 - 169 = 10,$

$S_6 = 26,$

$S_6 + DQ = 36,$

$TD_6 = TD_5 + S_6 + DQ = 159 + 36 = 195,$

$TE_6 = TE_5 + S_6 = 179 + 26 = 205.$

DETERMINISTIC QUEUEING SITUATIONS

For uninterrupted and uniform-flow situations no delay results if $y$(or $\rho$) $\leqslant 1$. However, there are many other uniform-flow situations of considerable interest and of relevance to a variety of traffic problems—particularly in the area of road traffic. The main ones are:

(a) interrupted service, with $y \leqslant 1$—the simple traffic-signal situation;

(b) transient interruption of service (an emergency situation arising from temporary reduction in capacity);

(c) peak overloading;

(d) cyclic variations in input or output or both.

### *Interrupted Service*

Consider a fixed-time traffic-light installation.

Let $c$ be the cycle time,

$a$ the proportion effective green time, i.e. the ratio of the green time to the total cycle time,

$q$ the arrival flow in a given approach,

$s$ the saturation flow at the start line of a given approach,

$y$ the degree of saturation, i.e. $q/sa$.

For stability of operation $y \leqslant 1$.

Now if $q(t)$ and $s(t)$ be the intensity of the input and output flows respectively, then for the case of uniform input

$$q(t) = q$$

and for pulsed output for one complete cycle

$$\begin{aligned} s(t) &= 0 \qquad 0 < t < (1-a)c \\ &= s \qquad c > t > (1-a)c. \end{aligned}$$

Reference to Fig. 3.15 shows that the number in the system at time $t$ is given by

$$L = \int_0^t \{q(t) - s(t)\}\, dt,$$

$$L = \int_0^t (q-s)\, dt$$

$$= q\int_0^t dt - s\int_{(1-a)c}^{t} dt$$

$$= s(1-a)c + (q-s)t.$$

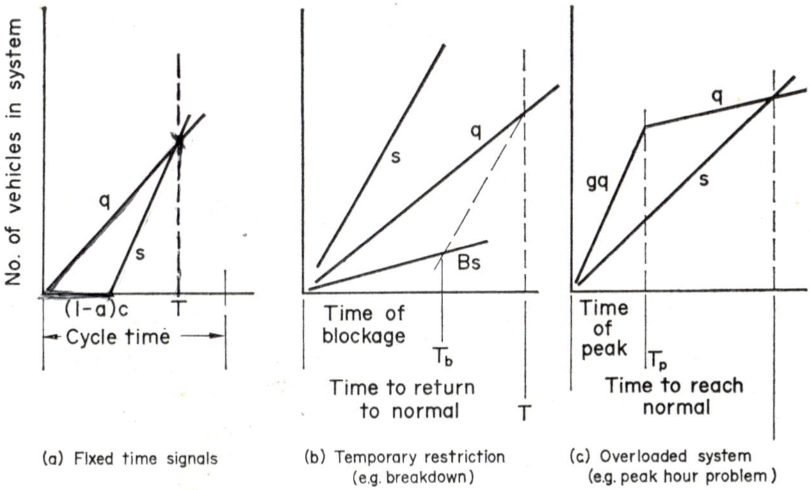


FIG. 3.15. Deterministic queueing situations.

The time $T$ to discharge the queue is thus

$$T = \frac{(1-a)c}{(1-ay)} \text{ (for } t = T \text{ when } L = 0). \qquad (3.25)$$

The total delay to all vehicles

$$D = \int_0^T (q-s)t\,dt$$

$$= q\int_0^T t\,dt - s\int_{c(1-a)}^{T} \{t-(1-a)c\}\,dt$$

$$= \frac{q(1-a)^2c^2}{2(1-ay)}. \tag{3.26}$$

The number of vehicles that arrive per cycle is $qc$. Therefore, the delay per vehicle (all vehicles in system)

$$\overline{W} = D/qc = \frac{(1-a)^2c}{2(1-ay)} \qquad ay \leqslant 1 \tag{3.27}$$

$$= \infty \qquad ay > 1.$$

This is sometimes known as Clayton's formula, and it was the first theoretical result used in intersection delay calculations.

### *Transient Reduction in Service—Breakdowns, Accidents, etc.*

An approach similar to that given above can be applied to the "emergency"-type situation arising from the closure of a lane or lanes due to accidents or maintenance; it can be applied also to the peak overload problem. Figure 3.15 illustrates the situation.

As before the input and output intensities are

$$q(t) = q,$$

$$s(t) = Bs \quad 0 < t \leqslant T_b \quad \text{and} \quad B < y$$

$$= s \qquad t > T_b$$

where $T_b$ is the time of the blockage.

Net number in system at time $t$

$$L = \int_0^t \{q(t)-s(t)\}\, dt$$

$$= \int_0^{T_b} (q-Bs)\, dt + \int_{T_b}^{t} (q-s)\, dt$$

$$= q\int_0^t dt - Bs\int_0^{T_b} dt - s\int_{T_b}^{t} dt$$

$$L = qt - BsT_b - sT_b - st$$

$$= (q-s)t + s(1-B)T_b.$$

If $T$ is the time till the effect of the blockage disappears, then $T = (t)_{L=0}$, i.e.

$$T = \frac{(1-B)}{s-q}\, sT_b = \frac{(1-B)T_b}{(1-y)}. \tag{3.28}$$

The total delay $D$ to all vehicles affected by the interruption is

$$D = \int_0^T \{q(t)-s(t)\}t\, dt$$

$$= q\int_0^T t\, dt - Bs\int_0^{T_b} t\, dt - s\int_{T_b}^{T} \{t-(1-B)T_b\}\, dt$$

$$= \frac{sT_b^2}{2(1-y)}(1-B)(y-B). \tag{3.29}$$

The average delay per vehicle is $D$/number of vehicles delayed. The number delayed is clearly $qT = q(1-B)T_b/(1-y)$,

$$\overline{W} = \frac{1}{2y}(y-B)T_b. \tag{3.30}$$

EXAMPLE

Consider a two-lane tunnel in which a breakdown occurs during the busy period. Capacity of each lane is 1250 vehicles per hour. If the busy-period flow level is 2000 vehicles per hour determine the length of time the resulting disturbance will take to disappear; delay to all vehicles and the number affected if the breakdown closes one lane for 20 minutes.

Here $s = 2500$ vehicles per hour,

$q = 2000$ vehicles per hour,

$y = 0{\cdot}8$,

$B = 0{\cdot}5$.

$$T = \frac{(1-B)}{(1-y)} T_b$$

$$= \frac{1-0{\cdot}5}{1-0{\cdot}8} \times 20 \text{ min}$$

$$= 50 \text{ min}$$

$$\overline{W} = \frac{1}{2y}(y-B)T_b = \frac{1}{2\times 0{\cdot}8}(0{\cdot}8-0{\cdot}5)\,20 = \frac{0{\cdot}3}{1{\cdot}6}\times 20$$

$$= 225 \text{ sec}$$

$$= 3 \text{ min } 45 \text{ sec.}$$

The number affected is given by

$$= q \times T = \frac{2000}{60} \times 50 = 1670.$$

***Temporary Overloading***

In the case of peak overload, the problem becomes one of considering the effects of variable $q$. If the peak overload factor is $g$ and the duration of the peak $T_p$, then as indicated in Fig. 3.15(c)

$$q(t) = gq \qquad 0 < t \leqslant T_p$$

$$= q \qquad t > T_p$$

$$s(t) = s \text{ for all } t.$$

Then the number in the system at time $t$ is

$$L = \int_0^{T_p} (gq-s)\,dt + \int_{T_p}^{t} (q-s)\,dt$$

$$= (gq-s)T_p + (q-s)(t-T_p).$$

The duration of the peak period effects is obtained when $L = 0$,

or

$$T = \frac{T_p(g-1)y}{(1-y)}. \qquad (3.31)$$

The total delay $D$ is

$$D = \int_0^{T} \{q(t)-s(t)\}t\,dt$$

$$= \int_0^{T_p} (gq-s)t\,dt + \int_{T_p}^{T} \{q(g-1)T_p+qt\}\,dt - \int_0^{T} st\,dt$$

$$= \frac{qT_p^2(g-1)(yg-1)}{2(1-y)}. \qquad (3.32)$$

The delay per vehicle is obtained by dividing by number of arrivals (= number of departures, $sT$)

and

$$sT = \frac{qT_p(g-1)}{(1-y)}$$

$$\overline{W} = \frac{T_p(yg-1)}{2}.$$

## *Cyclic Variations in Input and Output*

Figure 3.16 illustrates the flow situation through one phase of a signal assumed to have a square wave output, i.e.

$$s(t) = a\,.\,s - \frac{s}{\pi}\sum\frac{1}{n}\sin 2n\pi(1-a)\cos\frac{2n\pi t}{c}$$

$$-\frac{s}{\pi}\sum\frac{1}{n}\{1-\cos 2n\pi(1-a)\}\sin\frac{2n\pi t}{c}$$

where $s$ is the saturation flow and $a$ is the effective green time ratio and

$$s(t) = s(m-a)c < t < mc, \qquad m = 1, 2, \ldots, g$$
$$= 0 \text{ elsewhere.}$$

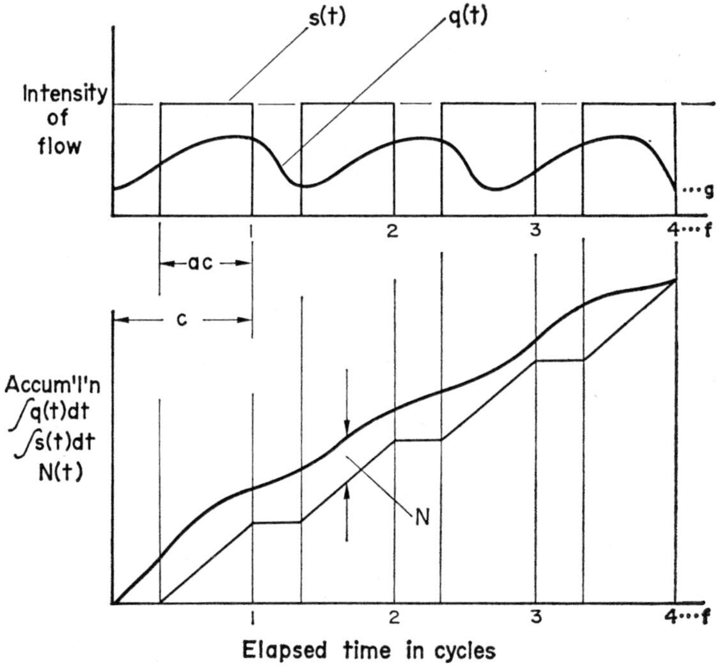


FIG. 3.16. Cyclic departure and arrival rates and accumulation on a given approach to a signalized intersection.

If now the input flow may be represented by a cyclic function of arbitrary form, viz.

$$q(t) = \sum A_n \cos \frac{2n\pi ft}{gc} + \sum B_n \sin \frac{2n\pi ft}{gc}$$

where

$$A_n = \frac{2f}{gc}\int_0^{gc/f} q(t)\cos\frac{2n\pi ft}{gc}\,dt$$

and

$$B_n = \frac{2f}{gc}\int_0^{gc/f} q(t)\sin\frac{2n\pi ft}{gc}\,dt$$

where $g$ full cycles of the input correspond to $f$ cycles of the signal phase—$f$ and $g$ have no common factor.

The delay per vehicle may be written

$$\overline{W} = \frac{1}{asgc}\int_0^{gc} N(t)\,dt$$

where $N(t)$ is the net number of vehicles that pass through in time $t$, i.e.

$$N(t) = \int_0^t \{q(t)-s(t)\}\,dt.$$

The formidable set of equations can be solved for the near saturation condition and leads to the surprisingly simple result

$$\overline{W} = \left[\frac{g}{2\pi asf}\sum\frac{1}{n}B_n+\tfrac{1}{2}(1-a)\right].c \tag{3.33}$$

The second term of this equation represents the delay inherent in the intersection itself, i.e. it is Clayton's result when $q = as$.

The first term takes account of the additional delay resulting from the phasing of the output from the previous intersection and the effect of different cycle time. In fact it provides the link between a given intersection approach and the one that feeds it. A more complete discussion of this approach to linked traffic signals is given in refs. 33 and 34.

## APPENDIX

## THE EQUATIONS OF STATE OF A MULTI-CHANNEL QUEUEING SYSTEM

A derivation of the equations of state of a queueing system with arbitrary service and arrival distributions is an esoteric exercise and the results achieved are generally intractable. When services and arrivals are of a Poisson kind the derivation is instructive and the results usable. The simplification results from the asymptotic property of the Poisson distribution, viz.

$$\underset{\delta t \to 0}{Lt}\, P_n(q\delta t) = \underset{t \to 0}{Lt}\, e^{-q\delta t}\frac{(q\delta t)^n}{n!} = 1 - q\delta t \quad \text{for} \quad n = 0$$

$$= q\delta t \qquad n = 1$$

$$= 0 \qquad n \geqslant 2.$$

Now for a queueing system with $M$ parallel channels each with service time $\overline{T}_s$ or service rate $1/\overline{T}_s$ two situations need to be distinguished;

(a) when the number $n$ in the system is less than $M$ the total service is proportional to $n$ the number in the system; in fact it equals $n/\overline{T}_s$,

(b) when $n$ is $\geqslant M$ then the service rate remains constant and equal to $M/\overline{T}_s$.

If the arrival rate is $q$ and there are $n$ units in the system at time $t$ then the probability that there will be still $n$ at time $t+\delta t$ is the sum of four independent probabilities:

(a) $P_n(t) \times$ Prob *no* arrivals $\times$ Prob *no* departures,

(b) $P_n(t) \times$ Prob *one* arrival $\times$ Prob *one* departure,

(c) $P_{n-1}(t) \times$ Prob *one* arrival $\times$ Prob *no* departures,

(d) $P_{n+1}(t) \times$ Prob *no* arrivals $\times$ Prob *one* departure.

Thus when $0 \leqslant n \leqslant M$

$$\begin{aligned} P_n(t+\delta t) = {} & P_n(t) \times (1 - q\delta t) \times \{1 - (n/\overline{T}_s)\delta t\} \\ & + P_n(t) \times q\delta t \times (n/\overline{T}_s)\delta t \\ & + P_{n-1}(t) \times q\delta t \times \{1 - ((n-1)/\overline{T}_s)\delta t\} \\ & + P_{n+1}(t) \times (1 - q\delta t) \times ((n+1)/\overline{T}_s)\delta t. \end{aligned}$$

Simplifying yields

$$\frac{dP_n(t)}{dt} = \frac{P_n(t+\delta t)-P_n(t)}{\delta t} = -(q+n/\overline{T}_s)P_n(t)+qP_{n-1}(t)$$
$$+\{(n+1)/\overline{T}_s\}P_{n+1}(t).$$

For steady-state conditions

$$\frac{dP_n(t)}{dt} = 0$$

and

$$qP_{n-1}+(n+1)/\overline{T}_sP_{n+1}-(q+n/\overline{T}_s)P_n = 0 \qquad \text{(i)}$$

when $n > M$ a new equation applies, viz.

$$qP_{n-1}+(M/\overline{T}_s)P_{n+1}-(q+M/\overline{T}_s)P_n = 0. \qquad \text{(ii)}$$

Further, there are special situations when $n = 0$ or $n = N$ if there is a limit $N$ on the number in the system. When $n = 0$ at time $t$, the conditions that will permit it to remain 0 at time $t+\delta t$ are

(a) $P_0(t) \times$ Prob *no* arrivals,
(b) $P_1(t) \times$ Prob *no* arrivals $\times$ Prob *one* departure.

Thus

$$P_0(t+\delta t) = P_0(t)\times(1-q\delta t)$$
$$+P_1(t)\times(1-q\delta t)\times(1/\overline{T}_s)\,\delta t$$

and for the steady-state situation

$$(1/\overline{T}_s)P_1-qP_0 = 0 \qquad \text{(iii)}$$

when $n = N$

$$(M/\overline{T}_s)P_N-qP_{N-1} = 0. \qquad \text{(iv)}$$

Starting with (iii) and substituting in (i) and then (ii) after $n = M$ one can establish the results given in (3.3). Thus

from (iii)

$$P_1 = (q\overline{T}_s)P_0 = AP_0.$$

Now (i) gives for $n = 1$

$$qP_0 + (2/\overline{T}_s)P_2 - \{q + (1/\overline{T}_s)\}P_1 = 0$$

and substituting (iii)

$$P_2 = \frac{q^2\overline{T}_s^2}{2}P_0 = \frac{A^2}{2}P_0.$$

Continuing up till $n = M-1$

$$P_n = \frac{A^n}{n!}P_0$$

when $n \geqslant M$

$$P_n = \frac{A^M}{M!}A^{(n-M)}P_0.$$

CHAPTER 4

# Mathematical Programming Methods

## INTRODUCTION

The object of this Chapter is to provide the student with some background understanding of and a little facility with linear programming and network flow algorithms. We have seen already how the land-use/transport system may be defined in network form and we have indicated how a mathematical programming approach may be directed to its optimization. Although the student no less than the practitioner will quickly pass his problem into the computer room, it is nevertheless essential for the student to have the experience of being able to work through some elementary and illustrative examples if he is to acquire a full measure of confidence with this new tool. In particular it is important to

(a) become familiar with certain basic definitions and propositions of linear algebra,

(b) be able to use the simplex method,

(c) understand the role and scope of the linear programming dual,

(d) have some knowledge of the maximum flow and shortest path algorithms of traffic networks.

There is now an extensive and rigorous literature on the subject.[38–42] The student will need to consult and study it if he is to acquire genuine facility with mathematical programming. It is not an easy literature for self-initiation, but it is hoped that this brief account will remove some of the strangeness of its ritualistic format.

## LINEAR ALGEBRA—DEFINITIONS AND BASIC PROPOSITIONS

Linear algebra is concerned with the properties and solution of linear equations and linear inequations or inequalities.

A system of $m$ linear inequations connecting $n$ variables may be written

$$\sum_{i=1}^{n} a_{ij}x_i \leqslant b_j \qquad j = 1, 2, \ldots, m, \tag{4.1}$$

i.e.

$$\begin{aligned} a_{11}x_1 + a_{21}x_2 + \ldots\ a_{n1}x_n &\leqslant b_1, \\ a_{12}x_1 + a_{22}x_2 + \ldots\ a_{n2}x_n &\leqslant b_2, \\ \cdot\quad \cdot\quad \cdot\quad \cdot\quad \cdot\quad &\cdot\quad \cdot \\ a_{1m}x_1 + a_{2m}x_2 + \ldots\ a_{nm}x_n &\leqslant b_m. \end{aligned} \tag{4.2}$$

TABLE 4.1.

| Var. | $b$ | $x_{n+1}$ | $x_{n+2}$ | ... | $x_{n+m}$ | $x_1$ | $x_2$ | ... | $x_n$ |
|---|---|---|---|---|---|---|---|---|---|
| $x_{n+1}$ | $b_1$ | 1 | | | | $a_{11}$ | $a_{21}$ | | $a_{n1}$ |
| $x_{n+2}$ | $b_2$ | | 1 | | | $a_{12}$ | $a_{22}$ | | $a_{n2}$ |
| | . | | | | | | | | |
| | . | | | | | | | | |
| | . | | | | | | | | |
| $x_{n+m}$ | $b_m$ | | | | 1 | $a_{1m}$ | $a_{2m}$ | | $a_{nm}$ |

Such a system of inequations may be readily converted to a system of equations by introducing $m$ slack variables,

i.e.

$$\sum_{i=1}^{n} a_{ij}x_i + x_{n+j} = b_j \qquad j = 1, 2, \ldots, m. \tag{4.3}$$

In what follows a "tableau" notation is common and (4.3) may be written as shown in Table 4.1.

There are now $m+n$ ($= N$, say) variables and a set of values $[x_i]$ ($i = 1, 2, \ldots, N$) which satisfied these equations is a *solution.* A feasible solution is a set of $x_i$ with no negative elements, i.e. the set $[x_i/x_i \geqslant 0]$.

BASIC SOLUTIONS

If any $m$ of the variables are chosen and the remaining ones set equal to zero then the $m$ equations in $m$ variables yield explicit sets $[x_i/i \in m, 0/i \in \bar{m}]$ which are known as basic solutions. The variables $x_i$, $i \in m$ are basic variables and those for $i \in \bar{m}$ non-basic. If the $x_i$ are all positive then they are basic feasible solutions. One should note, however, that the maximum number of basic solutions is $N!/\{m!(N-m!)\}$, i.e. ${}^nC_m$; but any set of equations may not necessarily yield this maximum for some of the $m \times m$ sets of equations may have a coefficient matrix which is singular.

The utility of linear algebra theory depends in large measure on two fundamental propositions:

(a) if a feasible solution exists then there is at least one basic feasible solution,

(b) unless a feasible solution is unbounded every feasible solution is the weighted average of the basic feasible solutions, i.e. the elements of $x_i$ may be written

$$x_i = t_i x_{i1} + t_2 x_{i2} + \dots t_K x_{iK} \qquad i = 1 \dots N \tag{4.4}$$

where

$$t_1 + t_2 + \dots t_K = 1 \quad \text{for} \quad t\text{'s} \geqslant 0$$

and

$$(x_{11}, x_{21}, \dots, x_{N1}), (x_{12}, x_{22}, \dots, x_{N2}), \dots (x_{1K}, x_{2K}, \dots x_{NK})$$

are basic feasible solutions.

AN EXAMPLE

The following example illustrates the major points set out in the foregoing review. Consider the set of inequalities

$$\begin{aligned} x_1 &\leqslant 8, \\ x_2 &\leqslant 4, \\ x_1 + 2x_2 &\leqslant 12. \end{aligned} \tag{i}$$

These may be readily converted to equations by adding three additional variables (i.e. slack variables), thus

$$\begin{aligned} x_1 \quad +x_3 \quad &= 8 \\ x_2 \quad +x_4 \quad &= 4 \\ x_1+2x_2 \quad +x_5 &= 12. \end{aligned} \qquad \text{(ii)}$$

There are five unknowns and three equations, hence $N = 5$, $m = 3$ and the maximum number of basic solutions is

$$N!/\{m!(N-m)!\} = 10.$$

TABLE 4.2.

| Basic variables | Non-basic | Solution | Vertex | Remarks |
|---|---|---|---|---|
| 1, 2, 3 | 4, 4 | (4, 4, 4, 0, 0) | *c* | Basic feasible |
| 1, 2, 4 | 3, 5 | (8, 2, 0, 2, 0) | *b* | ,, ,, |
| 1, 2, 5 | 3, 4 | (8, 4, 0, 0, −4) | *u* | Non-feasible |
| 1, 3, 4 | 2, 5 | (12, 0, −4, 4, 0) | *v* | ,, |
| 1, 3, 5 | 2, 4 | — | – | Singular |
| 1, 4, 5 | 2, 3 | (8, 0, 0, 4, 4) | *a* | Basic feasible |
| 2, 3, 4 | 1, 5 | (0, 6, 8, −2, 0) | *w* | Non-feasible |
| 2, 3, 5 | 1, 4 | (0, 4, 8, 0, 4) | *d* | Basic feasible |
| 2, 4, 5 | 1, 3 | — | | Singular |
| 3, 4, 5 | 1, 2 | (0, 0, 8, 4, 12) | 0 | Basic feasible |

The basic solutions can be obtained by solving (ii) in sets of three of the variables as summarized in Table 4.2. It can be seen that of the ten possible solutions five are basic feasible, three are non-feasible and two give singular solutions.

Figure 4.1 shows the solution in two-dimensional space. The hatched area is the feasible solution space and the basic feasible solutions are the points 0, *a*, *b*, *c*, *d*. It is clear from the diagram that *u*, *v* and *w* are non-feasible. In the algebraic solution the non-feasibility is apparent because of the negative elements that occur in respect of these points. The singularities are of course due to there being two pairs of parallel lines.

Consider now whether the solution (3, 3, 5, 1, 3) is feasible and if so express it in terms of the basic feasible solutions.

That it is feasible can be readily checked by direct substitution in (ii). More precisely its feasibility is established if it can be expressed as a

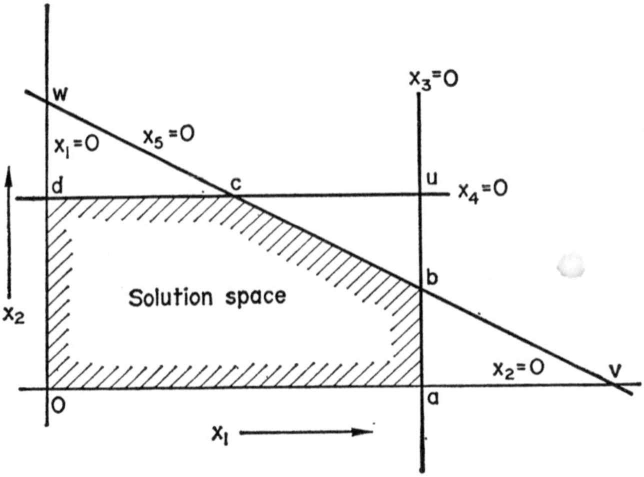


FIG. 4.1. Graphical description of the solution space for system of constraints in two variables.

linear combination of the basic solutions. This can in fact be done in this case, for

$$\begin{aligned}
8t_2+8t_3+4t_4 \qquad &= 3\\
2t_3+4t_4+4t_5 &= 3\\
8t_1 \qquad +4t_4+8t_5 &= 5\\
4t_1+4t_2+2t_3 \qquad &= 1\\
12t_1+4t_2 \qquad +4t_5 &= 3
\end{aligned}$$

give

$$t_1 = 0 \,.\, t_2 = \tfrac{1}{4} \,.\, t_3 = 0 \,.\, t_4 = \tfrac{1}{4} \quad \text{and} \quad t_5 = \tfrac{1}{2}$$

and

$$\sum t = 1.$$

## THE OPTIMIZING PROBLEM

In the simple example just treated the solution space was a plane polygon defined by the vertices 0, *a*, *b*, *c* and *d* and these in turn are the basic feasible solutions to equations of the system. For a three-dimensional problem the solution space would be the polyhedron formed by a set of planes and the basic solutions would represent the vertices formed by the enclosing planes. When the dimension of the problem is extended the basic feasible solutions are the extreme points of a convex region in *n*-dimensional space formed by the hyperplanes represented by the constraints of the system.

The optimizing problem is simply one of finding the extreme point that maximizes or minimizes the value of a given linear function of the variables which is itself a movable hyperplane. Two fundamental propositions that establish the validity of this intuitively plausible method are

(a) if there is an optimum solution then at least one basic feasible solution exists,

(b) provided the feasible region is convex (i.e. not unbounded) every feasible optimizing solution is a linear combination of basic feasible solutions.

## LINEAR PROGRAMMING

The exploitation of linear programming methodology started just on two decades ago with the pioneering work of George Dantzig who in 1947 was preoccupied with problems of military logistics and flow. Along with queueing theory and digital simulation it is one of the legs of the tripod on which supports the powerful methodology of operations research. There can be little doubt that two papers by Dantzig[43,44] published in 1951 mark the establishment of linear programming as a recognized mathematical calculus.

There is little need to review here its wide scope for the application of linear programming to economic production and organizational problems for we will have sufficient opportunity to demonstrate its possibilities in land-use and transport planning alone.

In formal terms the L.P. problem may be stated in the following terms:

If $x_1, x_2, \ldots, x_n$ are independent variables or activities, then a set $[x_i^0/x_i^0 \geqslant 0]$ may be chosen so as to maximize (or minimize) the linear function

$Z = \sum_{i=1}^{n} c_i x_i$ subject to a set of linear constraints, viz.

$$\sum_{i=1}^{n} a_{ij} x_i \leqslant b_j \qquad j = 1, 2, \ldots, m$$

the $a$'s, $b$'s and $c$'s may be positive, negative or zero.

The problem as stated above is generally described as the primal. There is, however, a meaningful reciprocal method of formulating it that is of fundamental importance to a full understanding of the methodology of linear programming and also of great practical utility. It may be stated

minimize (or maximize) the linear function

$$Y = \sum_{j=1}^{m} b_j y_j$$

subject to

$$\sum_{j=1}^{m} a_{ij} y_j \geqslant c_i \qquad i = 1, 2, \ldots, n.$$

For optimum sets $[x_i^0]$ and $[y_i^0]$ the values of the objective functions are equal, i.e. $Y = Z$.

The relationship between the variables in the primal and the dual may be usefully represented in tabular form (see Table 4.3).

The utility of the dual is both methodological and interpretative so far as the results are concerned. In the methodological aspect it may be more simple to solve the dual than the primal or vice versa. If the primal has many variables and few constraints then the dual will have few variables and numerous constraints and depending on other aspects of the format the solution of the one may be more easily accomplished than of the other. In addition, because of a special relationship between the primal and dual variables of the optimum solution a combined algorithm known as the primal-dual algorithm has great power of application to certain problems such as the Transportation Problem. This depends on the Complementary Slackness Theorem, viz.

> if the slack variable added to the *i*th constraint in the primal is different from zero then the *i*th variable of the dual will be zero in the optimum solution—and vice versa.

TABLE 4.3. PRIMAL-DUAL MATRIX

| D \ P | $x_1$ | $x_2$ | ... | $x_n$ | |
|---|---|---|---|---|---|
| $y_1$ | $a_{11}$ | $a_{21}$ | ... | $a_{nm}$ | $\leqq b_1$ |
| $y_2$ | $a_{12}$ | $a_{22}$ | ... | $a_{n2}$ | $\leqq b_2$ |
| . | | | | | |
| . | | | | | |
| . | | | | | |
| $y_m$ | $a_{1m}$ | $a_{2m}$ | | $a_{nm}$ | $\leqq b_m$ |
| | $\geqq c_1$ | $\geqq c_2$ | | $\geqq c_n$ | |

The interpretative utility depends on the fact that the dual variables are measures of the effect of incremental changes in the constraints on the objective function. In problems with an economic context they represent shadow prices or opportunity costs. In the desire-line problem the dual variables indicate the increase or decrease in the amount of travel, i.e. the transport task that will result from marginal changes in the land-use pattern.

## THE SIMPLEX METHOD

Although the Simplex Method of solving a linear programme is essentially an algorithm, it does nevertheless provide mathematical understanding of the theory of linear equations or inequations. Our concern here will be restricted to its role as an algorithm and in this role it is completely general in application. However, due to the rapid accumulation of variables and constraints in many practical problems—often 100's and sometimes 1000's—the computational effort involved in using it is beyond the scope of all but the biggest computers. There has been therefore a great deal of vigorous mathematical research into the development of modified and alternative algorithms for solving linear programming problems, particularly those with a special structure such as the transportation problem.

We will illustrate most of the important characteristics of the simplex method with an example.

As the motor car has no doubt been the major stimulus to the new science of land-use and transport planning, it is perhaps most appropriate in this context to take as an example one relating to the production of motor vehicles. In order to make use of the example already taken to demonstrate the basic notions of linear algebra we will be constrained to consider an extremely simple production problem.

TABLE 4.4.

| Department | Capacity (units/year) | |
|---|---|---|
| | Cars | Trucks |
| Manufacturing | | |
| components and sub-assemblies | 12,000 | 6000 |
| Production lines | | |
| cars | 8000 | |
| trucks | | 4000 |

AN EXAMPLE

Consider a motor-car plant in which the manufacturing facilities are divided broadly into production of components and sub-assemblies which feed two main production lines—one for cars, one for trucks. The capacities of the various units are shown in Table 4.4.

If the profit on cars is \$200 and on trucks \$100, what is the optimum production plan?

Assuming that the optimizing criterion is maximization of profit, the linear function is

$$Z = 200x_1 + 100x_2$$

where $x_1$ and $x_2$ are the number of cars and trucks to be produced.

The constraints imposed by production capacities may be written

$$x_1 \leqslant 8000,$$
$$x_2 \leqslant 4000,$$
$$x_1 + 2x_2 \leqslant 12{,}000.$$

If we introduce slack variables $x_3$, $x_4$, $x_5$ and rewrite the constraint equations in tableau form we have the results shown in Table 4.5.

TABLE 4.5.

| $c$ | | | 0 | 0 | 0 | 2 | 1 |
|---|---|---|---|---|---|---|---|
| | Variable | $b$ | $x_3$ | $x_4$ | $x_5$ | $x_1$ | $x_2$ |
| 0 | $x_3$ | 8000 | 1 | | | 1 | |
| 0 | $x_4$ | 4000 | | 1 | | | 1 |
| 0 | $x_5$ | 12,000 | | | 1 | 1 | 2 |

It will be noted that a new row and column have been added and these associate the coefficients of the linear function with their respective variables.

Table 4.5 immediately provides an initial solution—as it always does in cases where positive slack variables can be introduced into each of the constraint equations. The initial solution is $x_3 = 8000$, $x_4 = 4000$, $x_5 = 12{,}000$ and $x_1 = 0$, $x_2 = 0$. The solution at any stage of the subsequent calculations is given by the current values in the $b$ column for the corresponding variables in the basis, i.e. the elements of the Variable column. All other variables not occurring in the basis have the value 0.

The determination of the optimum solution now proceeds by constructing a succession of tableaux according to a simple set of rules. This is not an iterative method for the optimum solution will be reached in a finite number of steps. Each set of $[x_i]$ is an extreme point of the convex space defined by the constraint hyperplanes, in this case straight lines. It should be pointed out that it is in general not necessary to examine all extreme points. The method permits a relatively "direct" path to the optimum $[x_i]$ and a test at the completion of each new tableau indicates whether the optimum has been reached. The steps in Simplex algorithm are:

Step I.* Construct the initial tableau as indicated.

Step II. Form row $Z$ by summing the product of $c$ in the first column and the corresponding element of the columns $b$, $x_3$, $x_4$, $x_5$, $x_1$, $x_2$.

Step III. Form row $(Z-c)$ by subtracting $c$ taken from the top row.

Step IV. Examine row $(Z-c)$ and select the column variable with the largest negative value of $(Z-c)$—in this case $x_1$. Where two or more columns have equal maximum negative values select any convenient one.

Step V. Divide each positive element of the column variable selected into the corresponding element of the $b$ column. Note the smallest of these and replace the corresponding basis variable by the column variable obtained in Step IV. In this case $8000/1 < 12000/1$, $\therefore$ $x_3$ is removed and replaced by $x_1$. If none is positive the problem has an unbounded solution.

Step VI. Calculate new values of all the elements of the tableau starting with those in the row just introduced—here $x_1$. New values of the $x$-row elements are obtained simply by dividing the values in the previous tableau by the pivot element, i.e. the element common to the column $x_1$ and the old row $x_3$. All other row elements are obtained by

(i) taking old values (previous tableau),

(ii) subtracting the product $a \times b$ where

$a$ = element common to the column variable just introduced and the row currently being recalculated,

$b$ = corresponding element in the new basic variable.

In this example row $x_4$ is rewritten directly from the previous tableau because $a = 0$. For row $x_5$ we see in Table 4.6.

* The application of the various steps of the Simplex algorithm to the problem under consideration is shown in Table 4.8.

Step VII. Repeat steps II to VI until no negative values of $(Z-c)$ remain. In this example the third tableau yields positive values of all $(Z-c)$ and the solution is thus

$$x_1 = 8000,$$
$$x_2 = 2000,$$
$$x_3 = 0,$$
$$x_4 = 2000 \text{ (surplus capacity in truck assembly line)},$$
$$x_5 = 0,$$
$$\text{and } Z = \$1{\cdot}8 \times 10_6.$$

TABLE 4.6.

| Old value (from 1st tableau) | – | $a \times b$ | = | New value |
|---|---|---|---|---|
| 12,000 | – | $1 \times 8000$ | = | 4000 |
| 0 | – | $1 \times 1$ | = | $-1$ |
| 0 | – | $1 \times 0$ | = | 0 |
| 1 | – | $1 \times 0$ | = | 1 |
| 1 | – | $1 \times 0$ | = | 1 |
| 2 | – | $1 \times 0$ | = | 2 |

If we now superimpose on the solution space diagram of Fig. 4.1 the line $Z = 2x_1 + x_2$ we note that the extreme point that yields the optimum solution, i.e. maximum value of $Z$ is the point (8, 2, 0, 2, 0) 000's as shown in Fig. 4.2.

It is instructive now to consider the dual for it illustrates certain other essential details of the Simplex algorithm. The dual may be constructed as indicated in Table 4.3 above. As there were three constraints in the primal, three dual variables are required (see Table 4.7).

The dual equations are thus

$$y_1 \quad + y_3 \geqslant 200,$$
$$y_2 + 2y_3 \geqslant 100$$

and the objective function $Y = 8000y_1 + 4000y_2 + 12{,}000y_3$ is to be minimized.

Slack variables may now be introduced giving

$$\begin{aligned} y_1 \quad + y_3 - y_4 \quad &= 200, \\ y_2 + 2y_3 \quad - y_5 &= 100. \end{aligned}$$

It will be noted that they have negative signs and thus cannot be used to form an initial basis. A method of complete generality—but one

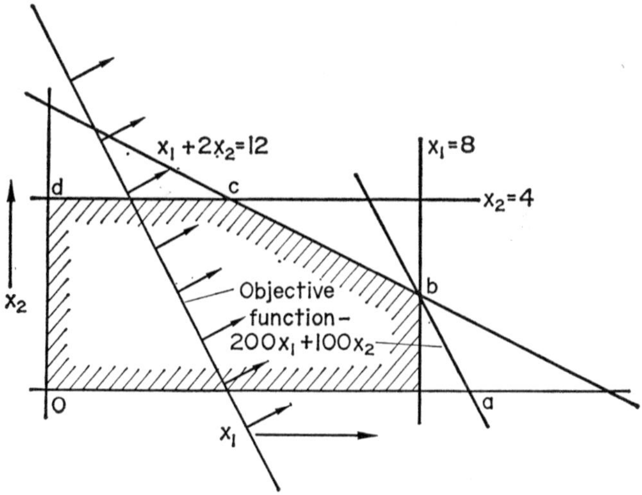


FIG. 4.2. Illustration of process of optimization of a linear function with linear constraints.

TABLE 4.7.

| P / D | $x_1$ | $x_2$ | $b$ |
|---|---|---|---|
| $y_1$ | 1 | | $\leqq$ 8000 |
| $y_2$ | | 1 | $\leqq$ 4000 |
| $y_3$ | 1 | 2 | $\leqq$ 12,000 |
| $c$ | $\geqq$ 200 | $\geqq$ 100 | |

which is unwieldy in large problems—is to introduce artificial variables $y_6$ and $y_7$, viz.

$$\begin{aligned} y_1 \quad + y_3 - y_4 \quad + y_6 \quad &= 200, \\ y_2 + 2y_3 \quad - y_5 \quad + y_7 &= 100, \end{aligned}$$

these variables are not required in the final solution and to ensure that they will not appear they have associated with them in the linear function large positive coefficients.

$y$ may be written in full as

$$Y = 8000y_1 + 4000y_2 + 12{,}000y_3 + 0y_4 + 0y_5 + My_6 + My_7.$$

TABLE 4.8. SIMPLEX TABLEAU FOR THE PRIMAL

| *C* | | | 0 | 0 | 0 | 2 | 1 |
|---|---|---|---|---|---|---|---|
| | Basic variable | *B* | $X_3$ | $X_4$ | $X_5$ | $X_1$ | $X_2$ |
| 0 | $X_3$ | 8000 | 1 | | | 1 | |
| 0 | $X_4$ | 4000 | | 1 | | | 1 |
| 0 | $X_5$ | 12,000 | | | 1 | 1 | 2 |
| *Z* | | | 0 | 0 | 0 | 0 | 0 |
| (*Z*−*C*) | | | | | | −2 | −1 |
| 2 | $X_1$ | 8000 | 1 | | | 1 | |
| 0 | $X_4$ | 4000 | | 1 | | | 1 |
| 0 | $X_5$ | 4000 | −1 | | 1 | 0 | 2 |
| *Z* | | | 2 | 0 | 0 | 2 | 0 |
| (*Z*−*C*) | | | 2 | 0 | 0 | 0 | −1 |
| 2 | $X_1$ | 8000 | 1 | | | 1 | |
| 0 | $X_4$ | 2000 | $\frac{1}{2}$ | 1 | $-\frac{1}{2}$ | | |
| 1 | $X_2$ | 2000 | $-\frac{1}{2}$ | | $\frac{1}{2}$ | 0 | 1 |
| *Z* | | | 3/2 | | $\frac{1}{2}$ | 2 | 1 |
| (*Z*−*C*) | | | 3/2 | 0 | $\frac{1}{2}$ | 0 | 0 |

TABLE 4.9. SIMPLEX TABLEAUX FOR THE DUAL

| $C$ | | | $-M$ | $-M$ | $-8$* | $-4$* | $-12$* | 0 | 0 |
|---|---|---|---|---|---|---|---|---|---|
| | Basic variable | $B$ | $y_6$ | $y_7$ | $y_1$ | $y_2$ | $y_3$ | $y_4$ | $y_5$ |
| $-M$ | $y_6$ | 200 | 1 | | 1 | | 1 | $-1$ | |
| $-M$ | $y_7$ | 100 | | 1 | | 1 | 2 | | $-1$ |
| $Y$ | | | $-M$ | $-M$ | $-M$ | $-M$ | $-3M$ | $+M$ | $+M$ |
| $(Y-C)$ | | | 0 | 0 | $-M+8$ | $-M+4$ | $-3M+12$ | $+M$ | $+M$ |
| $-M$ | $y_6$ | 150 | 1 | $-\frac{1}{2}$ | 1 | $-\frac{1}{2}$ | | $-1$ | $\frac{1}{2}$ |
| $-12$ | $y_3$ | 50 | | $\frac{1}{2}$ | | $\frac{1}{2}$ | 1 | | $-\frac{1}{2}$ |
| | | | $-M$ | $\frac{1}{2}M-6$ | $-M$ | $\frac{1}{2}M-6$ | $-12$ | $+M$ | $\frac{1}{2}M+6$ |
| $(Y-C)$ | | | $+M$ | $\frac{3}{2}M-6$ | $-M+8$ | $\frac{1}{2}M-2$ | 0 | $+M$ | $-\frac{1}{2}M+6$ |
| $-8$ | $y_1$ | 150 | 1 | $-\frac{1}{2}$ | 1 | $-\frac{1}{2}$ | | $-1$ | $\frac{1}{2}$ |
| $-12$ | $y_3$ | 50 | | $\frac{1}{2}$ | | $\frac{1}{2}$ | 1 | | $-\frac{1}{2}$ |
| $Y$ | | | $-8$ | $-2$ | $-8$ | $-2$ | $-12$ | $+8$ | $+2$ |
| $(Y-C)$ | | | $M-8$ | $M-2$ | 0 | $+2$ | 0 | 8 | 2 |

* 000's omitted.

The Simplex method can be applied without complication. But if we apply the rules as given above without change then $Y$ must be changed to $-Y$ so that the maximizing algorithm may be applied. Table 4.9 shows the full working of the Simplex method for the dual.

One should note first that the objective function is $\$1{\cdot}8\times10^6$ in both the primal and the dual. The values of $y$ in the optimum solution of the dual are

$$y_1 = 150, \qquad y_5 = 0,$$
$$y_2 = 0, \qquad y_6 = 0,$$
$$y_3 = 50, \qquad y_7 = 0.$$
$$y_4 = 0,$$

A number of observations are of interest. Firstly, there is a slight computational advantage with the dual. Although the number of tableaux is the same the effort required to recompute the elements is obviously much less.

The primal and dual solutions confirm the complementary slackness principle. For example, in the primal the second constraint reveals unused capacity in the truck production line and thus $y_2$ should be zero.

The economic interpretation is perhaps not at first obvious. $y_1$ corresponds to the first constraint of the primal. If one more unit of production capacity were available then one extra car could be made at the expense of half a truck, i.e.

$$1\times200-\tfrac{1}{2}\times100 = 150.$$

In other words, profit would be increased by \$150 (i.e. $y_1$).

## DEGENERACY

It would be remiss to conclude this brief introductory account of linear programming and the Simplex method without mentioning the problem of degeneracy. Mention, however, is all that is necessary, for whilst theoretically important, degeneracy is of little practical consequence to the user of L.P. algorithms. A degenerate solution of a linear system occurs when more than $N-m$ of the $X_i$'s in a basic solution have zero value. Although degenerate solutions are not in themselves invalid their occurrence in the Simplex method may lead to cycling and as a result an optimum solution may never be obtained. In practice, however, this possibility is so rare that computer programmes seldom, if ever, include statements to deal with it.

In the Simplex method degeneracy becomes imminent when the test for removal of a variable from the current basis fails to specify it uniquely. In other words, if there is a tie between or among two or more of the $b_j/a_{ij}$ quotients of step V of the algorithm then there will be at least one zero in the $b$ column of the ensuing tableau. In due course a situation may arise where a subsequent tableau will yield no reduction in the value of $Z$. If an arbitrary choice of the new basic variable is made, further iterations beyond this stage may produce a tableau identical with one already obtained. There are a number of ways of dealing with this situation.(38,39) One due to Charnes may be described as follows:

> If a tie occurs between a pair of $b_j/a_{ij}$ values then it may be resolved by successively comparing quotients $a_{rj}/a_{ij}$ where $r = 1, 2, \ldots n/r \neq i$ until tie is broken.

## APPLICATIONS TO LAND-USE/TRANSPORT PROBLEMS

The application of the programming approach in land-use/transport planning will be developed in full measure in appropriate sections of this text. However, at this point it is appropriate to outline the scope of its application, viz.

(a) shortest path problem,
(b) maximum flow problems,
(c) the desire-line problem—this is directly analogous to the "classical" L.P. Transportation Problem,
(d) desire-line pattern with capacity restraint,
(e) distribution of employment and population to land-use zones,
(f) assignment of traffic to a transport network,
(g) various combinations of (a), (b), (c), (d), (e) and (f),
(h) capital budgeting problems,
(i) organizational and operational problems.

Although the shortest-path and maximum-flow problem can be set up as linear programs, their special format permits their solution by simple combinational labelling algorithms. These are explained below.

## THE SHORTEST-PATH ALGORITHM

There is a number of good algorithms for determining the shortest path through a network. They have been described in a series of papers between 1958 and 1963 from a variety of independent workers in the U.S.A., Britain and Europe and it is difficult to be precise in ascribing priority to the various authors.

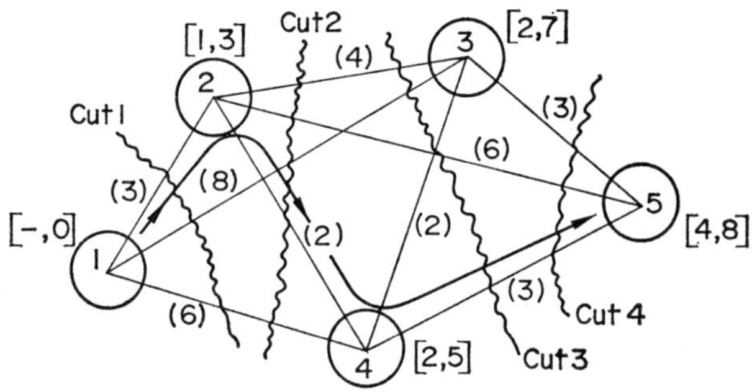


FIG. 4.3. Illustration of shortest-path algorithm showing cuts and labels.

The most elegant and perhaps most useful for hand calculators is the one described by Ford and Fulkerson.[45] For computer application the most economical is that due to Dykstra in Holland and Whiting and Hillier in Britain.[46] The former will best serve the student in his exploration of the scope and application of this methodological device.

Consider the network shown in Fig. 4.3. The numbers on the links represent their "length" in either distance or time units. In general our primary concern is with the shortest path between a given pair or a number of pairs of nodes which represent desire-line origins and destinations.

In this case let the desire lines on the network be (1, 5) and (1, 3). Our primary concern is then to find the shortest routes from 1 to 5 and 1, 3. However, the algorithm to be explained and most others find the shortest route from a given origin to all the nodes of the network as a matter of course.

The labelling algorithm consists of finding a "label"—a two-element set—for a node, $j$, which is the other end of the link $(i, j)$ which has emanated from an already labelled node, $i$. The label is written $[i, p(j)]$ where $i$ then is the node from which the current node $j$ is labelled, and $p(j)$ is the shortest path from the origin to $j$.

The origin may be clearly labelled $[-, 0]$ and the next node to be labelled is found by determining the $j$ for which

$$p(j) = \min_{i \in X} [p(i) + \min_{j \in \bar{X}} l(i, j)]$$

where $X$ and $\bar{X}$ are the subsets of the universal set of nodes $N$ which lie to either side of the section (known as a cut) which divides the labelled and the unlabelled nodes at any stage. The process is repeated until $\bar{X}$ contains no elements.

In the above example we start with Cut 1. The only $i$ in $X$ is node 1. The possible $j$'s in $\bar{X}$ are 2, 3 and 4. $p(1) = 0$ and

$$\min_{j=2,\,3 \text{ or } 4} [l(1, 2), l(1, 3) \text{ or } l(1, 4)],$$

i.e. min (3, 8 or 6) is 3, i.e. $l(1, 2)$.

Therefore $j = 2$ and $p(2) = 3$.

Then node 2 is labelled [1, 3].

Cut 2 is now defined and $i$ may equal 1 or 2

$$p(j) = \min_{i=1,2} \begin{bmatrix} p(1) + \min\limits_{j=3,4} (l(1, 3), l(1, 4)) \\ p(2) + \min\limits_{j=3,4,5} (l(2, 3), l(2, 4), l(2, 5)) \end{bmatrix}$$

$$= \min \begin{bmatrix} 0 + \min (8, 6) \\ 3 + \min (4, 2, 6) \end{bmatrix}$$

which is satisfied when $p(j) = 5$ giving $j = 4$

$$p(4) = p(2)+l(2, 4).$$

Then node 4 is labelled [2, 5].

Proceeding in this way gives labels for the other nodes as shown. The shortest path to node 5 is 8 units in length and the links comprising it are determined by travelling back to the origin guided by the first element of the label. In this case 5 → 4 then 4 → 2 and finally 2 → 1. The shortest path to node 3 is 7 units.

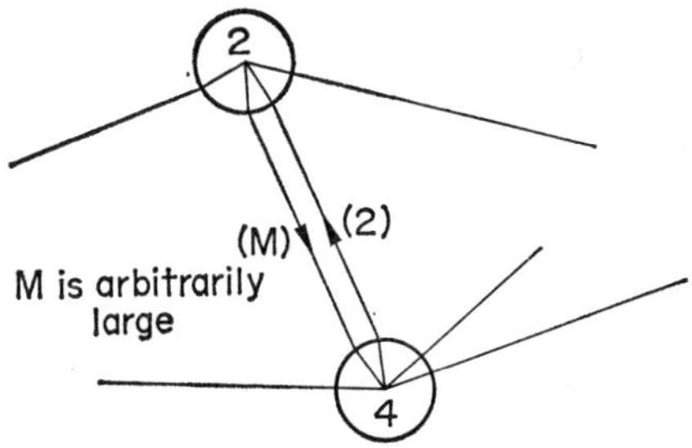


FIG. 4.4. Method of representing a one-way link.

In the problem just solved no restriction has been placed on the direction in which any of the links may be traversed. However, one-way streets may be dealt with by replacing the one-way link or links with a pair and giving to one member of the pair an arbitrarily large length or travel time—say $M$. If link (4, 2) were one way from 4 to 2 then it would appear as shown in Fig. 4.4.

If the network had been solved with this restraint then the shortest path from 1 to 5 would have been either 1, 2, 5 or 1, 4, 5 and its length 9 units.

### *Less-short Paths*

The problem of finding the second, third, ..., $n$th shortest path is of practical importance in assignment and certain other traffic problems. However, there are difficulties in defining alternative paths in a network because the concept of an alternative path ranges all the way from an alternative path for any particular link of the shortest path to one which has no links whatsoever in common with it. Problems also arise in connection with loops and two-way travel on a given link. For example, in the problem dealt with above the realistic second shortest path from 1 to 2 is via 4 and its length 8 units. However, theoretically—provided the link (2, 3) is undirected—it could be 1 to 2, 2, 4 and 4, 2, i.e. 7 units. It is also apparent that the number of alternative paths in any sizeable network soon becomes unmanageably large.

There are a number of algorithms that assist with the development of alternative paths. All of them require the shortest path to be firstly determined. The most conceptually general algorithm is one due to Bellman and Kalaba.[47] An algorithm proposed by Hoffman and Pavley is useful for single *o–d* pairs.[48]

If $p_1(j)$ is the shortest path to node $j$, then the $n$th shortest path $p_n(j)$ is

$$p_n(j) = \min \begin{bmatrix} \min_{1\, i\in I} [p_n(i)+l(i,j)] \\ \min_{2\, i\in I} [p_{n-1}(i)+l(i,j)] \\ \cdot\ \cdot\ \cdot\ \cdot\ \cdot\ \cdot\ \cdot\ \cdot\ \cdot \\ \min_{n-1\, i\in I} [(p_2(i)+l(i,j)] \\ \min_{n\, i\in I} [(p_1(i)+l(i,j)] \end{bmatrix}$$

For the second shortest path

$$p_2(j) = \min \begin{bmatrix} \min_{1\, i\in I} [p_2(i)+l(i,j)] \\ \min_{2\, i\in I} [p_1(i)+l(i,j)] \end{bmatrix}$$

where $\min_n [p_1, p_2, p_3, \ldots, p_n]$ is defined as the $n$th lowest value of the set of path lengths $p$. $I$ is the set of "origin" nodes joined to $j$.

Because of the above-mentioned difficulties associated with looping and retracing of steps it is helpful to apply the $n$th shortest-path algorithms to directed networks. This can be done without any serious loss of generality.

If we take the network of Fig. 4.3 and direct the links as shown in Fig. 4.5 the shortest path labels remain the same.

The second shortest path labels may now be developed. With node 1 as origin there is clearly no alternative path from node 1 to 2 and it may thus be labelled $[-, M]$ where $M$ is arbitrarily large.

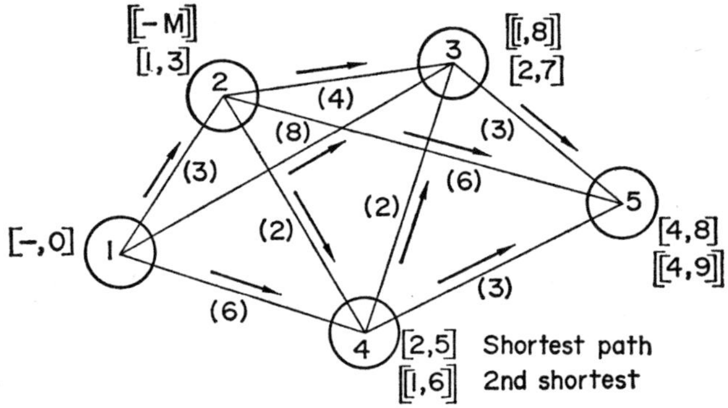


FIG. 4.5. Application of the Bellman–Kalaba algorithm to second shortest path; second shortest path labels are double bracketed.

Consider now node 4.

$$p_2(4) = \min \begin{bmatrix} \min\limits_{i=1,2}{}_1 \begin{bmatrix} p_2(1)+l(1,4) \\ p_2(2)+l(2,4) \end{bmatrix} \\ \min\limits_{i=1,2}{}_2 \begin{bmatrix} p_1(1)+l(1,4) \\ p_2(2)+l(2,4) \end{bmatrix} \end{bmatrix} = \min \begin{bmatrix} \min_1 \begin{bmatrix} M+6 \\ M+2 \end{bmatrix} \\ \min_2 \begin{bmatrix} 0+6 \\ 3+2 \end{bmatrix} \end{bmatrix}$$

$$= \min \begin{bmatrix} M+2 \\ 6 \end{bmatrix} = 6.$$

The second shortest path to node 4 is therefore 6 and it arises from $l(1, 4)$ and is therefore the path $(1, 4)$.

For node 3.

$$p_2(3) = \min \begin{bmatrix} \min\limits_{i=1,2,4}{}_1 \begin{bmatrix} p_2(1)+l(1,3) \\ p_2(2)+l(2,3) \\ p_2(4)+l(4,3) \end{bmatrix} \\ \\ \min\limits_{i=1,2,4}{}_2 \begin{bmatrix} p_1(1)+l(1,3) \\ p_1(2)+l(2,3) \\ p_1(4)+l(4,3) \end{bmatrix} \end{bmatrix} = \min \begin{bmatrix} \min_1 \begin{bmatrix} M+8 \\ M+4 \\ 6+2 \end{bmatrix} \\ \\ \min_2 \begin{bmatrix} 0+8 \\ 3+4 \\ 5+2 \end{bmatrix} \end{bmatrix}$$

$$= \min \begin{bmatrix} 8 \\ 8 \end{bmatrix} = 8.$$

There are thus two equal "seconds" to node 3—(1, 3) and (1, 4) (4, 3).

For node 5.

$$p_2(5) = \min \begin{bmatrix} \min\limits_{i=2,3,4}{}_1 \begin{bmatrix} p_2(2)+l(2,5) \\ p_2(3)+l(3,5) \\ p_2(4)+l(4,5) \end{bmatrix} \\ \\ \min\limits_{i=2,3,4}{}_2 \begin{bmatrix} p_1(2)+l(2,5) \\ p_1(3)+l(3,5) \\ p_1(4)+l(4,5) \end{bmatrix} \end{bmatrix} = \min \begin{bmatrix} \min_1 \begin{bmatrix} M+6 \\ 8+3 \\ 6+3 \end{bmatrix} \\ \\ \min_2 \begin{bmatrix} 3+6 \\ 8+3 \\ 5+3 \end{bmatrix} \end{bmatrix}$$

$$= \min \begin{bmatrix} 9 \\ 9 \end{bmatrix} = 9.$$

Here again there are two equal second shortest paths.

## MAXIMUM FLOW THROUGH A NETWORK

It is axiomatic that the maximum flow through a network comprised of a single chain of links (elements) in series is equal to the saturation flow of the smallest of its elements. It is also axiomatic that the maximum flow of a simple parallel arrangement of links is equal to the sum of their saturation flows. However, in a generalized series/parallel arrangement of traffic-flow elements the maximum flow is by no means intuitively obvious. Although the elegant theory of graphs introduces a comprehensive and novel algebra it is helpful to the student with engineering background to recognize the close links between generalized activity networks and those of electrical and hydraulic engineering.

Consider the network shown in Fig. 4.5, the links of which have capacities as indicated by the numbers written against them and flows as designated by $q(i,j)$.

Nodes 1 and 5 may be regarded variously and respectively as origin and destination, source and sink, input and output, positive and negative terminals. The intermediate nodes 2, 3 and 4 are branch points or junctions. The familiar concept of conservation of flow gives

$$\begin{aligned} q(o, N)-q(N, o) &= Q && i = o, \quad \text{i.e. } 1, \\ q(i, N)-q(N, i) &= 0 && i \neq o, d, \text{ i.e. } 2, 3, 4, \\ q(d, N)-q(N, d) &= -Q && i = d \quad \text{i.e. } 5, \end{aligned}$$

where $q(i,j)$ is the flow in link $(i,j)$—from $i$ to $j$,

$q(j, i)$ is the reverse flow,

$N$ is the set of nodes that make up the network,

$Q$ is the total flow through the network,

and $q(i, N) = \sum_{j \in N} q(i,j)$.

The capacity constraints yield

$$q(i,j) \leqslant s(i,j) \qquad \text{for all links } (i,j).$$

### *A Section (or Cut)*

A section is a line which cuts all paths between the origin and destination. Such a section divides the set of nodes, $N$, into two subsets, $X$ and $\bar{X}$. If $Q$ is the flow through the network then for any section

$$\sum_{\substack{i \in X \\ j \in \bar{X}}} q(i,j) - \sum_{\substack{i \in X \\ j \in \bar{X}}} q(j, i) = Q \leqslant \sum_{\substack{i \in X \\ j \in \bar{X}}} s(i,j).$$

The maximum flow through the network is therefore set by the section with the minimum saturation flow. The problem of determining it is to find the section $(X, \bar{X})$ that minimizes

$$\sum_{\substack{i \in X \\ j \in \bar{X}}} s(i,j).$$

Although the problem can be set up as a linear programme the number of constraints makes the taking out of a solution, even with computer help, "uneconomic". In simple and small networks a solution may often be obtained by inspection. In general, however, the sophisticated but elegant labelling algorithm first described by Ford and Falkerson provides the student with a most instructive and useful tool.

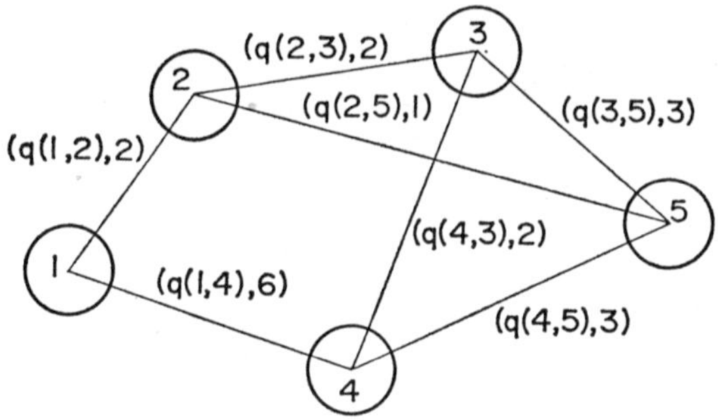


FIG. 4.6. Basic network with link capacities defined.

If we define the residual flow of an element or link,

$$r(i,j) = s(i,j) - q(i,j),$$

the algorithm may proceed in two stages.

1. An improved solution may be obtained from any feasible solution by loading any path between the origin and destination with a flow equal to the min $r(i,j)$ for all links $(i,j)$ of the path being loaded. It is instructive for the traffic analyst to thus load the network by successively loading the shortest and second, third ... paths for this will often represent a "practical" maximum flow.

For example, in Fig. 4.6 let it be assumed that (1, 2, 3, 5), (1, 4, 5) and (1, 4, 3, 5) are respectively the first, second and third shortest paths, then

path (1, 2) (2, 3) (3, 5) has min $r(i,j) = 2$,

(1, 4) (4, 5) has min $r(i,j) = 3$,

and (1, 4) (4, 3) (3, 5) has min $r(i,j) = 1$.

It should be noted that min $r(ij)$ are obtained successively. A feasible solution to the flow status of the network is thus that

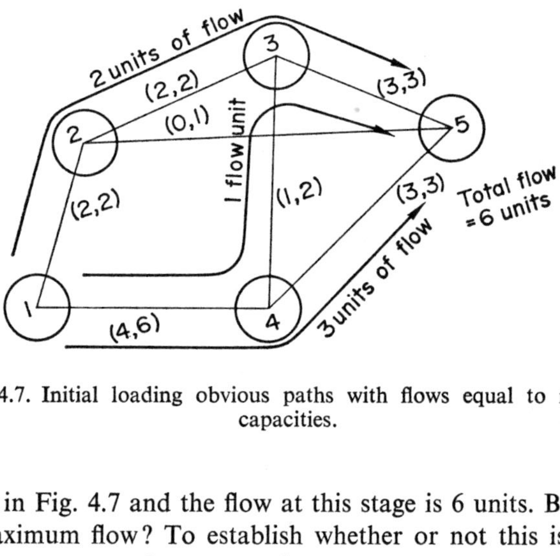


FIG. 4.7. Initial loading obvious paths with flows equal to minimal capacities.

shown in Fig. 4.7 and the flow at this stage is 6 units. But is this the maximum flow? To establish whether or not this is so it is necessary to proceed to a second stage.

2. It may be possible by rearranging the flow pattern to increase it. For example, an extra unit of flow may reach node 5 via link (2, 5). But if this is to be done then the flow in (2, 3) must be reduced by one unit and consequently in (3, 5). At first this may appear to restore the *status quo*. In fact, however, the reduction of flow in (3, 5) may be compensated by drawing an additional flow of one

unit from node 1 via (1, 4), (4, 3) and (3, 5). Although obvious in this case, it is not generally so. The Ford and Falkerson algorithm will, however, always resolve the situation. The label is a three element set

$$[\pm, i, d(j)]$$

The label is attached to any node $j$ joined to $i$. If the flow is $i$ to $j$ the first element is $+$; if from $j$ to $i$ the first element is $-$. The second element is the origin end of the link $(i, j)$ and $d(j)$ is

$$\min\,[d(i), r(i,j)] \qquad \text{provided } r(i,j) > 0$$

or

$$\min\,[d(j), q(j,i)] \qquad \text{if} \qquad q(j,i) > 0.$$

If $r(i,j) = 0$, $j$ cannot be labelled from $i$.

The labelling process is continued until the destination is labelled $d(d)$. The flow increment $d(d)$ is then added or subtracted respectively to the elements of the chain traced from the destination back to the origin according to whether the first element of the label is $+$ or $-$. The process is repeated until it is impossible to reach the destination. The section (or cut) dividing the labelled and unlabelled nodes is the minimum capacity section.

In the example of Fig. 4 we may proceed from the solution already obtained by loading the various paths as follows.

The origin is labelled $(+, o, M)$—the $+$ indicates flow is always out of the origin, the $o$ indicates it is generated at the origin at an arbitrarily large magnitude $M$.

Proceeding from node 1. Node 2 cannot be labelled from 1 as $r(1, 2) = 0$. Node 4 can be labelled for $r(1, 4) = 2$, i.e. $> 0$. Its label is $(+, 1, 2)$ as $r(1, 4) < d(1)$, i.e. $< M$.

From node 4, node 5 cannot be labelled but 3 can be labelled $(+, 4, 1)$ because $r(4, 3)$, i.e. 1 is less than $d(4)$ which is 2.

From node 3, node 5 cannot be labelled but 2 can be labelled $(-, 3, 1)$ because $d(3) < q(3, 2)$, i.e. $1 < 2$.

From node 2 the only unlabelled node is 5. It can be labelled $(+, 2, 1)$.

The last element of this label (refer Fig. 4.8(a)), i.e. 1 is now superimposed on the existing flows to yield the pattern shown in Fig. 4.8(b).

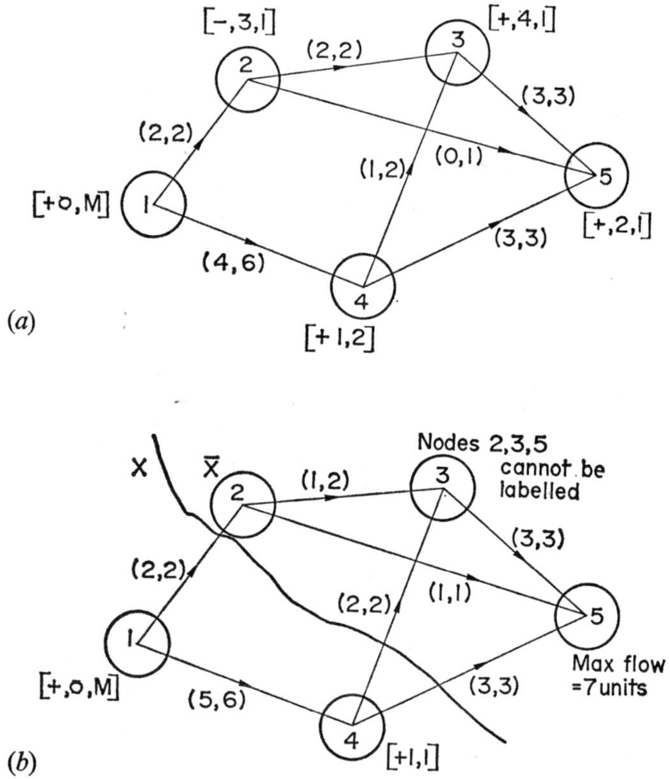


**Fig. 4.8.** Application of Ford–Fulkerson algorithm to initial solution shown in Fig. 4.7.

Repeating the process shows that only nodes 1 and 4 can be labelled Fig. 4.8(b). Hence the flow is maximum and the minimum section is as shown, i.e. $X = [1, 4]$, $\bar{X} = [2, 3, 5]$. In this example there are subsequent sections of similar capacity but $(X, \bar{X})$ represents the first section to be fully saturated.

One could have started the labelling procedure with all flows in the network elements set equal to zero. More iterations would have been required, of course, before the optimum solution was obtained.

## THE TRANSPORTATION PROBLEM

Because of the central place of the "Transportation Problem" in the general development of linear programming theory and because of the scope for its application to land-use and transport planning problems we will consider here the desire-lines and population-distribution problems.

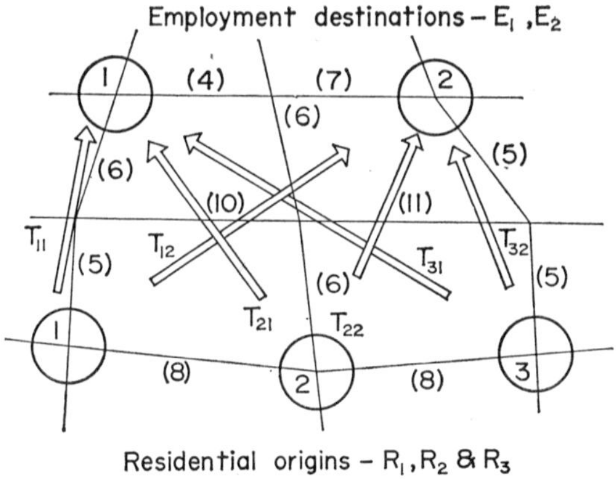


FIG. 4.9. The application of classical transportation problem to desire-line determination.

Figure 4.9 describes a land-use/transport system with two work zones and three residential zones. The zone centres are connected by a set of arterial roads with link travel times as shown in brackets. The work zones (destination) $d = 1, 2$ have 4000 and 5000 jobs respectively and the residential zones $o = 1, 2, 3$ each contain 3000 workers. Determine the desire-line pattern that will minimize the total amount of work trip travel in the system.

The minimum travel times between the origin/destination pairs may be obtained by inspection (the student may verify them by using the shortest-path algorithm) and are shown in Table 4.10. The only constraints in this problem are the trip end constraints, viz.

$$\begin{aligned}
T_{11}+T_{21}+T_{31} &= E_1 = 4,\\
T_{12}+T_{22}+T_{32} &= E_2 = 5,\\
T_{11}+T_{12} &= R_1 = 3,\\
T_{21}+T_{22} &= R_2 = 3,\\
T_{31}+T_{32} &= R_3 = 3.
\end{aligned}$$

TABLE 4.10. INTERZONAL TRAVEL TIMES

| *o* \ *d* | 1 | 2 |
|---|---|---|
| 1 | 11 | 22 |
| 2 | 16 | 18 |
| 3 | 21 | 10 |

Further, $\sum E = \sum R$, and because of this any one of the above constraints is redundant—in this case say the last one. The objective function is the total travel time and may be expressed

$$Z = 11T_{11}+16T_{21}+21T_{31}+22T_{12}+18T_{22}+10T_{32}.$$

Because the constraints are equations it is necessary to introduce artificial $T_1 \ldots T_4$ variables with their associated $M$'s in the objective function. Table 4.11 gives the initial and final tableaux of the Simplex solution.

The desire-line volumes are thus

$$\begin{aligned}
T_{11} &= 3000 \qquad & T_{12} &= 0\\
T_{21} &= 1000 \qquad & T_{22} &= 2000\\
T_{31} &= 0 \qquad & T_{32} &= 3000
\end{aligned}$$

and the total amount of travel = 115,000 person minutes.

TABLE 4.11. SIMPLEX TABLEAUX (INITIAL AND FINAL)

| $C$ | | | $-M$ | $-M$ | $-M$ | $-M$ | $-11$ | $-16$ | $-21$ | $-22$ | $-18$ | $-10$ |
|---|---|---|---|---|---|---|---|---|---|---|---|---|
| | Var. | $E$'s, $R$'s | $T_1$ | $T_2$ | $T_3$ | $T_4$ | $T_{11}$ | $T_{21}$ | $T_{31}$ | $T_{12}$ | $T_{22}$ | $T_{32}$ |
| $-M$ | $T_1$ | 4000 | 1 | | | | 1 | 1 | 1 | | | |
| $-M$ | $T_2$ | 5000 | | 1 | | | | | | 1 | 1 | 1 |
| $-M$ | $T_3$ | 3000 | | | 1 | | 1 | | | 1 | | |
| $-M$ | $T_4$ | 3000 | | | | 1 | | 1 | | | 1 | |
| $Z$ | | | $-M$ | $-M$ | $-M$ | $-M$ | $-2M$ | $-2M$ | $-M$ | $-2M$ | $-2M$ | $-M$ |
| $Z-C$ | | | — | — | — | — | $\begin{cases}-2M\\+11\end{cases}$ | $-2M$ $+16$ | $-M$ $+21$ | $-2M$ $+21$ | $-2M$ $+18$ | $-M$ $+10$ |
| $-16$ | $T_{21}$ | 1000 | 1 | | $-1$ | | | 1 | 1 | $-1$ | | |
| $-10$ | $T_{32}$ | 3000 | 1 | 1 | $-1$ | $-1$ | | | 1 | | | 1 |
| $-11$ | $T_{11}$ | 3000 | | | 1 | | 1 | | | 1 | | |
| $-18$ | $T_{22}$ | 2000 | $-1$ | | 1 | 1 | | | $-1$ | 1 | 1 | |
| $Z$ | | | $-8$ | $-10$ | $-3$ | $-8$ | $-11$ | $-16$ | $-8$ | $-13$ | $-18$ | $-10$ |
| $Z-C$ | | | $\begin{cases}-8\\+M\end{cases}$ | $-10$ $+M$ | $-3$ $+M$ | $-8$ $+M$ | — | — | 13 | 9 | — | — |

The dual may now be set up by reference to the Primal/Dual Variable Matrix (see Table 4.12).

TABLE 4.12.

| D \ P | $T_{11}$ | $T_{21}$ | $T_{31}$ | $T_{12}$ | $T_{22}$ | $T_{32}$ | E's and R's |
|---|---|---|---|---|---|---|---|
| $D_1$ | 1 | 1 | 1 | | | | 4000 |
| $D_2$ | | | | 1 | 1 | 1 | 5000 |
| $O_1$ | 1 | | | 1 | | | 3000 |
| $O_2$ | | 1 | | | 1 | | 3000 |
| $O_3$ | | | 1 | | | 1 | 3000 |
| $t$'s | 11 | 16 | 21 | 22 | 18 | 10 | |

Note: the redundant equation of the primal has been included so that every $O$ and $D$ will be represented explicitly by a dual variable.

The constraint equations and the objective function of the dual may be written down, viz.

$$D_1+O_1 \leqslant 11$$
$$D_1+O_2 \leqslant 16$$
$$D_1+O_3 \leqslant 21$$
$$D_2+O_1 \leqslant 22$$
$$D_2+O_2 \leqslant 18$$
$$D_2+O_3 \leqslant 10.$$

It should be noted that because the constraints of the primal in this example are equalities, the signs of the dual variables are not restricted. Ample discussion on the mathematical aspects of this matter are contained in all standard texts on Linear Programming. We will not discuss it further here as it does not introduce complications.

The objective function is

$$W = 4000D_1+5000D_2+3000O_1+3000O_2+3000O_3.$$

The student will find no difficulty in verifying by the Simplex algorithm that

$$
\begin{aligned}
D_1 &= 8 & y_1 &= 0\\
D_2 &= 10 & y_2 &= 0\\
O_1 &= 3 & y_3 &= 13\\
O_2 &= 8 & y_4 &= 9\\
O_3 &= 0 & y_5 &= 0\\
 & & y_6 &= 0
\end{aligned}
$$

where $y_1$, $y_2 \ldots y_6$ are the slack variables introduced to form the initial basis.

The value of the objective function can be readily seen to be 115,000 person minutes, which of course agrees with that of the primal.

TABLE 4.13. MARGINAL TRAVEL MATRIX*

| | $E_1$ | $E_2$ | $R_1$ | $R_2$ | $R_3$ |
|---|---|---|---|---|---|
| $E_1$ | 0 | +2 | −5 | 0 | −8 |
| $E_2$ | −2 | 0 | −7 | −2 | −10 |
| $R_1$ | 5 | 7 | 0 | 5 | −3 |
| $R_2$ | 0 | 2 | −5 | 0 | −8 |
| $R_3$ | 8 | 10 | 3 | 8 | 0 |

* Transfer of trip ends between *R*'s and *E*'s are considered inappropriate.

## THE MARGINAL TRAVEL MATRIX

The dual variables permit a table to be constructed (Table 4.13) showing the marginal *increases* in the primal objective function due to unit change in the trip ends.

The elements of Table 4.13 are simply differences between the values of the dual variables corresponding to the *O*'s or *D*'s. For example, the element $R_1$, $R_2$ is obtained from dual variables $O_1 = 3$, $O_2 = 8$.

Having constructed this table it is possible to determine readily that the shift of a job from $E_1$ to $E_2$ would increase the amount of travel in the system by 2 units or moving a resident from zone 1 to 3 would reduce the travel by 3 units. Such marginal changes often set off a train of readjustment of trips. If one relates the travel time matrix to the diagram of Fig. 4.9 it will be clear that in the first case the transfer of a job from $E_1$ to $E_2$ results in one less trip on $T_{21}$ at 16 units and one more on $T_{23}$ at 18 units. On the other hand, moving a resident from $R_1$ to $R_3$ calls for contingent moves. One less trip on $T_{11}$ saves 11 units. However, if the surplus job at $E_1$ were made up by a trip from the new location of the residential end at $R_3$ then 21 travel units would be added, i.e. a net increase in travel of 10 units. Instead, however, a trip from $R_2$ fills the vacant job at a cost of 16 units. This means one less from $R_2$ to $E_2$ and thus a saving of 18 units. The surplus job at $E_2$ is then filled from $R_3$ at a cost of 10 units. Therefore the net change is $-11+16-18+10 = -3$, i.e. the value shown in the marginal travel matrix.

Further discussion on this important matrix will be given in the Chapter dealing with land-use models.

## THE PRIMAL-DUAL ALGORITHM

Because of the special format of the transportation problem the relationship between the primal and dual permits it to be solved as a network flow problem. Before proceeding to illustrate the Primal-Dual Network Algorithm it is appropriate to generalize the formulation of the primal and dual of the transportation problem, viz.

For the Primal

$$\text{minimize} \quad Z = \sum_o \sum_d t_{od} T_{od},$$

$$\text{subject to} \quad \sum_o T_{od} = R_o \qquad o = 1, 2 \dots m,$$

$$\sum_d T_{od} = E_d \qquad d = 1, 2 \dots n,$$

$$\text{and} \quad \sum_o R_o = \sum_d E_d$$

$$\text{and} \quad T_{od} \geqslant 0, \qquad R_o \geqslant 0, \qquad E_d \geqslant 0,$$

and for the Dual

$$\text{maximize} \quad Y = \sum_o O_o R_o + \sum D_d E_d,$$

$$\text{subject to} \quad O_o + D_d \leqslant t_{od},$$

where the dual variables $O_o$ and $D_d$ are unrestricted in sign, and $R_o > 0$, $E_d > 0$.

Now we already know that for optimal values

$$Z = Y.$$

Therefore

$$\sum_o \sum_d t_{od} T_{od} = \sum_o O_o R_o + \sum_d D_d E_d.$$

But

$$R_o = \sum_d T^*_{od} \quad \text{and} \quad E_d = \sum_o T^*_{od}.$$

Therefore

$$\sum_o \sum_d t_{od} T^*_{od} = \sum_o O^*_o \sum_d T^*_{od} + \sum_d D^*_d \sum_o T^*_{od}$$

or

$$\sum_o \sum_d T^*_{od}[t_{od} - (O^*_o + D^*_d)] = 0$$

and hence the interesting and useful result of the complementary slackness theorem:

if

$$T^*_{od} \neq 0 \quad \text{then} \quad O^*_o + D^*_d = t_{od}$$

and

$$T^*_{od} = 0 \quad \text{then} \quad O^*_o + D^*_d \neq t_{od}.$$

These relationships provide the basic rationale of the primal-dual network algorithm.

### Example

Consider again the example of the previous section. The $3 \times 2$ origin and destination network may be represented as a network as shown in Fig. 4.10.

A master origin is connected to the trip origins with links having capacities equal to their respective trip ends. Similarly with respect of the master destination $D$.

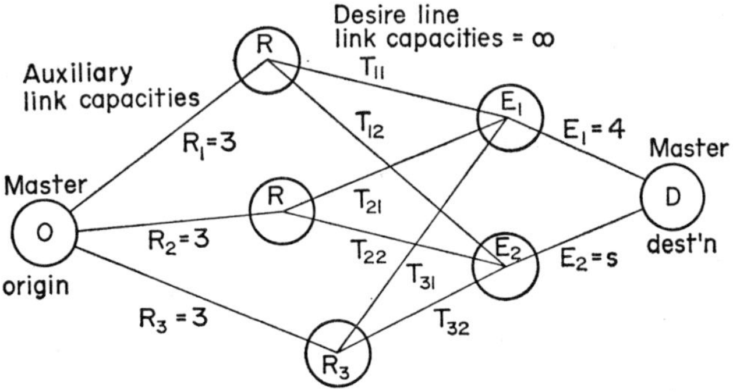


FIG. 4.10. Representation of the desire-line problem as a flow network.

The primal-dual algorithm starts by taking a feasible solution of the dual and then determining by the network flow algorithm a corresponding solution of the primal. This procedure is iterated until the maximum flow balances the total trip ends, i.e. until

$$\sum R_o = Q_{\max} = \sum E_d.$$

The constraints of the dual may be represented

| $d$ / $o$ | $D_1$ | $D_2$ |
|---|---|---|
| $O_1$ | 11 | 22 |
| $O_2$ | 16 | 18 |
| $O_3$ | 21 | 10 |

If $D_1$ and $D_2$ are set equal to zero and $O_1$, $O_2$, $O_3$ to the min $(t_{od})$ for $d = 1, 2$ respectively, then clearly we have a feasible solution.

Thus

$$D_1 = 0, D_2 = 0, O_1 = 11, O_2 = 16, O_3 = 10$$

and

| | |
|---|---|
| $O_1 + D_1 = 11 = t_{11}$ | therefore $T_{11}$ exists, |
| $O_1 + D_2 = 11 < t_{12}$ | ,, $T_{12}$ is zero, |
| $O_2 + D_1 = 16 = t_{21}$ | ,, $T_{21}$ exists, |
| $O_2 + D_2 = 16 < t_{22}$ | ,, $T_{22}$ is zero, |
| $O_3 + D_1 = 10 < t_{31}$ | ,, $T_{31}$ is zero, |
| $O_3 + D_2 = 10 = t_{32}$ | ,, $T_{32}$ exists. |

We can thus draw a network.

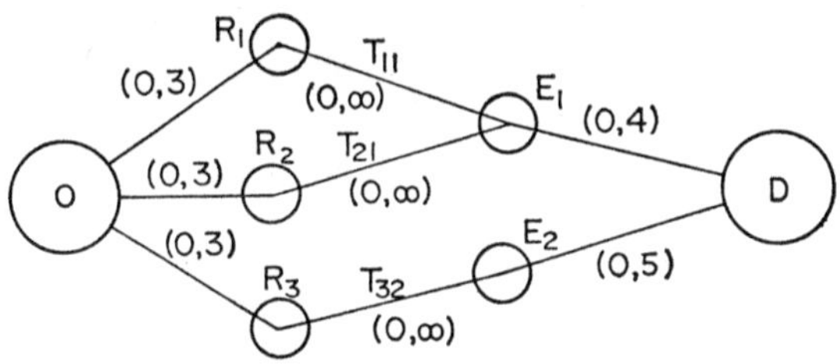


The capacity of the links $OR_1$, $OR_2$, $OR_3$ and $E_1D$, $E_2D$ are equal to the trip-end constraints set by $R_1$, $R_2$, $R_3$, $E_1$ and $E_2$ respectively. The capacity of links $T_{11}$, $T_{21}$ and $T_{32}$ are not constrained.

The maximum flow algorithm may be applied and the third iteration gives the maximum flow condition, viz.

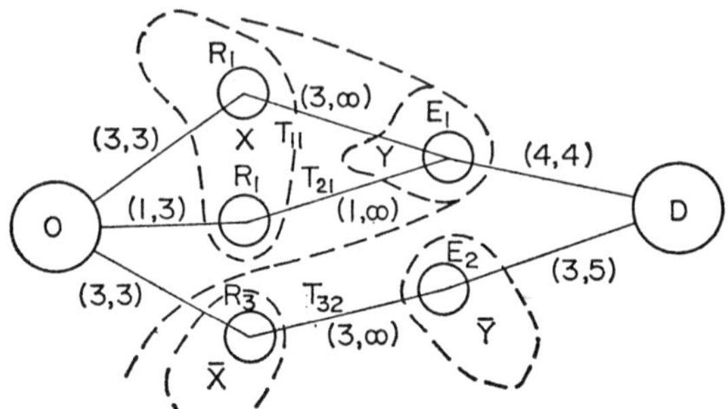

The maximum flow is 7 but this is less than $\sum R = \sum E = 9$. Therefore a further iteration is necessary.

The broken line section encloses to the left all labelled nodes.

It is appropriate now to construct the Dual-Variable Matrix embodying the initial solution, i.e.

DUAL-VARIABLE MATRIX ($O_o + D_d$)

| | | | $Y$ | $\bar{Y}$ |
|---|---|---|---|---|
| | | $D_d$ | 0 | 0 |
| | $O_o$ | $o \backslash d$ | 1 | 2 |
| $X$ | 11 | 1 | 11 | 11 |
| $X$ | 16 | 2 | 16 | 16 |
| $\bar{X}$ | 10 | 3 | 10 | 10 |

In order to increase the flow in the $O$–$D$ network it is necessary to bring in a new link because the stage just completed has determined the maximum flow possible with the initial set of $T_{od}$.

If we divide the labelled and unlabelled origins and destinations of the first solution into sets $X$, $\bar{X}$ and $Y$, $\bar{Y}$ then a feasible set of new links must have origins in $X$ and destinations in $\bar{Y}$. In order to determine which $T_{od}$ it is necessary to determine $O_o + D_d = t_{od}$ for $o \in X$ and $d \in \bar{Y}$.

An inspection of the initial solution indicates that $O_2 + D_2$ may be increased from 16 to 18 by increasing $O_2$ by 2 and hence $T_{22}$ may be introduced into the network. However, if this is done then the other equalities containing $O_2$ must be reduced by 2, i.e. their $D$ values must be reduced by this increment.

This process may be effected systematically by subtracting the $X\bar{Y}$ sub-matrix of the Dual-Variable Matrix from the corresponding elements of the travel time matrix and selecting the smallest element of the array for adjustment of the $O$'s in $X$ and the $D$'s. In this case the $X\bar{Y}$ sub-matrix is

$$\begin{bmatrix} 11 \\ 16 \end{bmatrix}.$$

The corresponding sub-matrix of the travel time table is

$$\begin{bmatrix} 22 \\ 18 \end{bmatrix}.$$

The min element after subtraction is 2.

The new dual-variable matrix may be written thus

| | $D_d$ | −2 | 0 |
|---|---|---|---|
| $O_o$ | o \ d | 1 | 2 |
| 13 | 1 | 11 | 13 |
| 18 | 2 | 16 | 18 |
| 10 | 3 | 8 | 10 |

and the new set of constraints may be written

$$\begin{aligned}
O_1+D_1 &= 11 = t_{11}, \\
O_1+D_2 &= 13 < t_{1}\ , \\
O_2+D_1 &= 16 = t_{21}, \\
O_2+D_2 &= 18 = t_{22}, \\
O_3+D_1 &= \ \ 8 < t_{31}, \\
O_3+D_2 &= 10 = t_{32}.
\end{aligned}$$

The new network thus comprises $T_{11}$, $T_{21}$, $T_{32}$ and in addition $T_{22}$, viz.

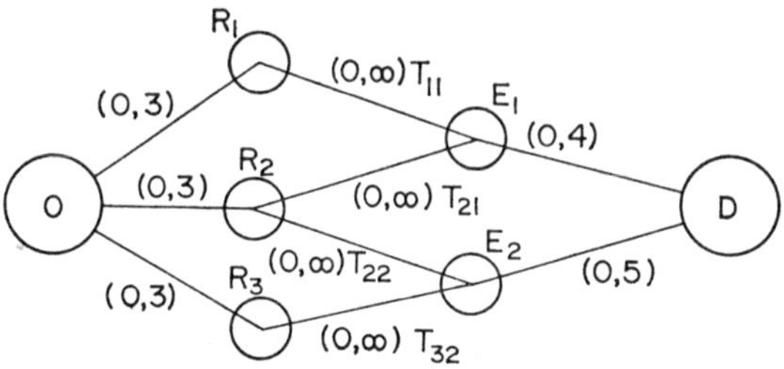

Application of the network flow algorithm gives after 5 steps

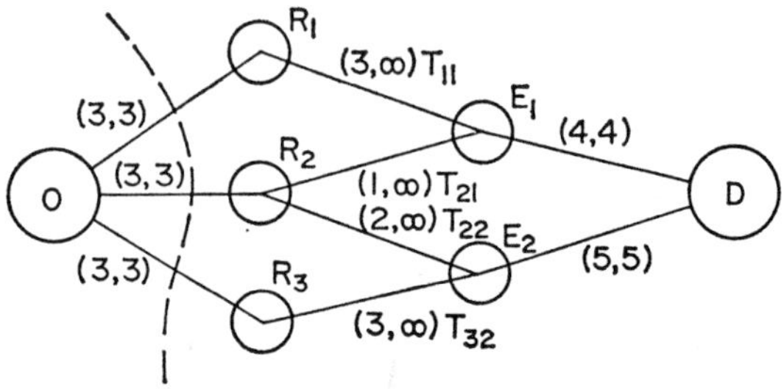


The max flow $= 9 = \sum R$, $\therefore$ algorithm complete.
The solution is thus

$$T_{11} = 3 \quad T_{21} = 1 \quad T_{31} = 0$$
$$T_{12} = 0 \quad T_{22} = 2 \quad T_{32} = 3$$

CAPACITY CONSTRAINTS ON DESIRE LINES

If capacity limitations $S_{od}$ are imposed by the transport network on the various desire lines, $od$, then the additional constraints $T_{od} \leqslant S_{od}$ in the primal give rise to new dual variables, $P_{od}$, such that the dual constraints become

$$t_{od} - O_o - D_d - P_{od} \geqslant 0 \qquad o = 1, 2, \dots O,$$
$$d = 1, 2, \dots D.$$

The algorithm explained above may be used, with minor modification to solve the "capacitated" transport problem. It is necessary to define sets of $od$-pairs as follows:

$A = [od \in OD/P_{od} < 0]$, i.e. a set of saturated links at any stage of the algorithm.

$B = [od \in OD/(t_{od} - O_o - D_d - P_{od} = 0), (P_{od} = 0)]$, i.e. links with flow but not as yet saturated.

$C = A \cup B$ the set of links included in the network at any stage.

The algorithm may proceed starting with an initial network given by

$$O_o = \min_d t_{od},$$
$$D_d = \min_o (t_{od} - O_o),$$
$$P_{od} = 0.$$

After determining the maximum flow further networks are defined if $Q < \sum O_o$ (or $\sum D_d$). Such networks are determined by a dual-variable matrix defined in terms of the labelled and unlabelled subsets $X$, $\bar{X}$ and $Y$, $\bar{Y}$ identified at the conclusion of each maximum flow solution, viz.

$$O'_o = \begin{cases} O_o + k & \text{for } o \in X \\ O_o & o \in \bar{X}, \end{cases}$$

$$D'_d = \begin{cases} D_d - k & \text{for } d \in Y \\ D_d & d \in \bar{Y}, \end{cases}$$

$$P'_{od} = \begin{cases} P_{od} - k & \text{for } od \in X\bar{Y} \cap C \\ P_{od} + k & od \in \bar{X}Y \cap A \\ P_{od} & \text{otherwise,} \end{cases}$$

where

$$k = \min \left[ \min_{X\bar{Y} \cap \bar{C}} (t_{od} - O_o - D_d - P_{od}), \min_{\bar{X}Y \cap A} |P_{od}| \right].$$

If in the previous example the capacity limits on the desire lines are

$$S_{11} = 1000 \text{ trips/unit time}$$
$$S_{12} = 3000 \quad ,, \quad ,, \quad ,,$$
$$S_{21} = 2000 \quad ,, \quad ,, \quad ,,$$
$$S_{22} = 2000 \quad ,, \quad ,, \quad ,,$$
$$S_{31} = 3000 \quad ,, \quad ,, \quad ,,$$
$$S_{32} = 2000 \quad ,, \quad ,, \quad ,,$$

then the optimum distribution of the traffic desires becomes

$$T_{11} = 1000 \qquad T_{21} = 2000 \qquad T_{31} = 1000$$
$$T_{12} = 2000 \qquad T_{22} = 1000 \qquad T_{32} = 2000$$

The resulting travel is now 146,000 person minutes—a considerable increase on that obtained when no capacity restrictions are imposed (i.e. 115,000 person minutes).

Verification of this result is left to the student.

CHAPTER 5

# Transport System Characteristics and Performance

## INTRODUCTION—COMPONENTS OF A TRANSPORT SYSTEM

We have already considered in the abstract sense the basic law of traffic flow which relates the travel time, delay or cost to the traffic intensity. We turn now to the study of the physical characteristics of transport systems in order that we may be able to

(a) give meaningful numerical values to the parameters and constants of the traffic flow law;
(b) define and calculate the output of the transport system;
(c) develop systems parameters and indices which will permit comparisons to be made between one mode and another and between major parts of a given mode.

The generalized transport system is made up of five components known as modes and a variety of terminal and transfer facilities that permit it to be loaded with passengers and freight and to permit the transfer from one mode to another.

The five modes are road, rail, sea, air and walking (or manhandling). The terminal facilities are ports, airports, railway and bus stations, marshalling yards and road haulage terminals. However, if walking and manhandling are considered as a mode then parking stations, loading docks, kerbside loading zones, bus stops, baggage counters and ticket windows may be properly included in the general definition terminal facilities.

It should be pointed out that the flow of passengers and freight through the terminal facilities is governed by the basic delay/flow law and such quantities as bus-loading times, ship-turnaround times are all components of the total time of the journey.

It is appropriate to consider too that all transport modes have two major components—the fixed facilities and the vehicle. In a superficial way one tends to the view that sea and air transport are not constrained in the same way as the other modes, i.e. by their rails and pavements to constrain the whole of the journey and as a result suffer extra delay because of the finite capacity of the way. But when we pause to consider that a modern jet aircraft can be delayed 10–30 minutes in a holding pattern after a 50-minute flight from an origin 500 miles away it is clear that the constraints of the fixed facilities have overall application.

We will consider here the following characteristics of the transport system:

(a) route factors of transport networks;
(b) the saturation-flow characteristics of various transport system elements and modes;
(c) the zero-flow travel times, service rates and free speeds;
(d) operating costs;
(e) terminals and transfer facilities;
(f) transport system output.

## ROUTE FACTOR

The route factor is a useful measure of effectiveness of the geometrical characteristics of the road or rail network. It can also be used as a measure of comparative performance of different operational schemes, e.g. one-way streets, route scheduling.

### *Definition of the Route Factor*

It is a simple matter to define the route factor for a single transport link and this may be done in a number of ways—

$$\text{Distance Route Factor } (R_D) = \frac{\text{Actual distance between } O \text{ and } D}{\text{Airline or direct distance}}.$$

Time and cost route factors may be defined in a similar way, viz.

$$\frac{\text{Actual time or cost between } O \text{ and } D}{\text{Arbitrary time or cost}}.$$

The arbitrary specification of the denominator makes these latter forms somewhat nebulous. However, if we consider instead the notion of comparative route factors then all three versions are both meaningful and useful.

$$\text{Comparative Route Factor } (R_{1,2}) = \frac{\text{Distance, time or cost on route 1}}{\text{Distance, time or cost on route 2}}$$

TABLE 5.1. AIRLINE DISTANCES BETWEEN CAPITAL CITIES AND ROAD, RAIL AND SEA-ROUTE FACTORS[49]

| Origin–Destination | Airline distance (miles) | Route factors | | |
|---|---|---|---|---|
| | | Road | Rail | Sea |
| Sydney–Melbourne | 452 | 1·25 | 1·30 | 1·45 |
| Sydney–Adelaide | 737 | 1·21 | 1·46* | 1·53 |
| Sydney–Perth | 2000 | 1·3 | 1·35 | 1·21 |
| Sydney–Brisbane | 480 | 1·33 | 1·28 | 1·22 |
| Melbourne–Brisbane | 866 | 1·39 | 1·39 | 1·42 |
| Melbourne–Perth | 1826 | 1·20 | 1·16 | 1·04 |
| Melbourne–Adelaide | 418 | 1·10 | 1·15 | 1·39 |
| Adelaide–Perth | 1408 | 1·22 | 1·15 | 1·08 |
| Brisbane–Cairns | 902 | 1·39 | 1·16 | 1·03 |

* Via Melbourne—due to the unique Australian problem of different rail gauges in the various States.

For a single desire-line journey or corridor the route factor is a function of mode, locational geometry and traffic operations. Table 5.1 gives a summary of the distance route factor characteristics for the principal transport modes for intercity travel on the Australian continent.

### *Network Route Factors*

The route factor of a network may be defined as the ratio of the total distance travelled between every pair of nodes to the sum of direct distance between them. In many situations it is appropriate to consider the route factor property of the network with respect to a single origin node—the central business district for example. In such circumstances the network route factor may be written

$$R_N = \frac{d_{o1}R_{o1} + d_{o2}R_{o2} + \ldots d_{o(N-1)}R_{o(N-1)}}{d_{o1} + d_{o2} + \ldots d_{o(N-1)}} \tag{5.1a}$$

where $d_{oj}$ is the direct distance between the origin and node $j$,
$R_{oj}$ is the route factor of the shortest path from the origin to node $j$.

$R_N$ may be written

$$R_N = \frac{d_{o1}}{\sum d_{oj}} R_{o1} + \frac{d_{o2}}{\sum d_{oj}} R_{o2} \ldots \frac{d_{o(N-1)}}{\sum d_{oj}} R_{o(N-1)}, \tag{5.1b}$$

i.e. the weighted average of individual route factors—weighted in terms of trip length.

It may be noted that the individual route factors may be further weighted by the traffic flows between the origin and the network nodes. However, at this stage our concern is primarily directed at developing a figure of merit or a systems parameter that describes the "geometry" of the network or its operational mode.

### *Some Route Factor Properties of Simple Networks*

1. *The square grid*

Consider a square grid comprised of two sets of $n+1$ parallel streets as shown in Fig. 5.1. The total distance from the origin to all other nodes may be obtained by summing the paths from the origin to all nodes on a transverse diagonal,
i.e.

$$\begin{aligned} a[2\times1+3\times2+4\times3+\ldots\ (n+1)n \\ 1\times2n+2(2n-1) \qquad\qquad n(n+1)] \\ = n(n+1)^2a = \textstyle\sum d\,.\,R. \end{aligned}$$

Further there are $(n+1)^2-1$ nodes (other than the origin). Therefore the average length of a trip is

$$\frac{n(n+1)^2a}{(n+1)^2-1} = \frac{(n+1)^2a}{(n+2)}. \tag{5.2}$$

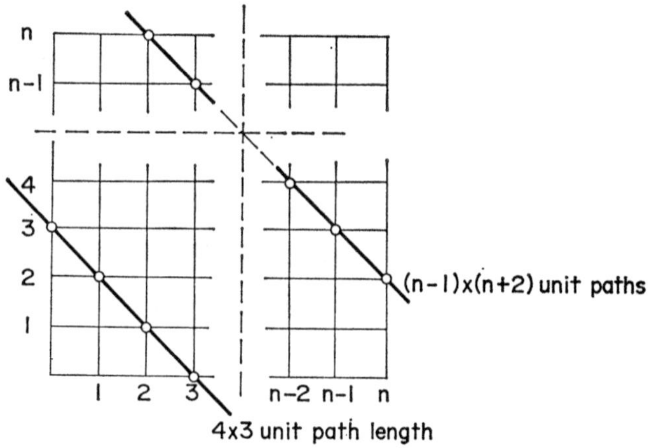


FIG. 5.1. An $n \times n$ square grid network.

The sum of the direct distances is

$$\begin{aligned} &2[1+2+3+\ldots n]a \\ &+\sqrt{2}[1+2+3+\ldots n]a \\ &+2[(2^2+1)^{\frac{1}{2}}+(3^2+1)^{\frac{1}{2}}+\ldots (n^2+1)^{\frac{1}{2}} \\ &\qquad +(3^2+2^2)^{\frac{1}{2}}+\ldots (n^2+2^2)^{\frac{1}{2}} \\ &\qquad\qquad \cdot\quad \cdot\quad \cdot\quad \cdot\quad \cdot \\ &\qquad\qquad \cdot\quad \cdot\quad \cdot\quad \cdot\quad \cdot \\ &\qquad\qquad\qquad \{n^2+(n-1)^2\}^{\frac{1}{2}}]a \end{aligned}$$

$$= n(n+1)\left[1+\frac{\sqrt{2}}{2}\right]a + a\sum_{r=2}^{n}\sum_{s=1}^{r-1}(r^2+s^2)^{\frac{1}{2}}.$$

The system route factor is thus

$$R_N = \frac{n(n+1)^2}{n(n+1)\left[1+\frac{\sqrt{2}}{2}\right]+2\sum_{r=2}^{n}\sum_{s=1}^{r-1}(r^2+s^2)^{\frac{1}{2}}}. \qquad (5.3)$$

The above expression, though clumsy, permits the systematic evaluation of the route factor of a square grid. However, it is tempting to seek a more simple expression. One that suggests itself is the simple average of the individual route factors. This is quite a plausible proposition if one considers the average route factor to arise from taking a simple random sample of nodes (by taking a random number pair) to represent the coordinates of any destination node. Thus

$$R_N \doteqdot R_A = \frac{1}{n}(R_1+R_2+\ldots R_n).$$

Table 5.2 compares values of $R_A$ and $R_N$ for square grids for $n = 2$ to 7.

TABLE 5.2. COMPARISON OF SYSTEM AND AVERAGE ROUTE FACTORS

| Grid size | $n$ | $R_A$ | $R_N$ |
|---|---|---|---|
| 2×2 | 2 | 1·19 | 1·22 |
| 3×3 | 3 | 1·21 | 1·24 |
| 4×4 | 4 | 1·22 | 1·25 |
| 5×5 | 5 | 1·23 | 1·26 |
| 6×6 | 6 | 1·24 | 1·27 |
| 7×7 | 7 | 1·25 | 1·27 |
| — | ∞ | 1·30 | 1·30 |

It should be noted that for a square grid system

$$R_{max} = 1{\cdot}41 \quad R_{min} = 1{\cdot}00 \quad R_{AV} = 1{\cdot}30 \text{ (for a large network).}$$

2. *Average distance from a point to points on a lineal route*

If $a$ is the perpendicular distance from one end of a line of length $b$ then the average distance $\bar{d}$ to all points on the line is

$$\bar{d} = \frac{1}{2b}\left[b\sqrt{(a^2+b^2)}+a^2\sinh^{-1}\frac{b}{a}\right]. \qquad (5.4)$$

3. *Average distance from the corner of a rectangle to internal points*

If the rectangle has sides of length $a$ and $b$ then

$$\bar{d} = \frac{1}{6ab}\left[a^3 \sinh^{-1}\frac{b}{a}+b^3 \sinh^{-1}\frac{a}{b}+2ab\sqrt{(a^2+b^2)}\right]. \quad (5.5)$$

Each of these results can be used to show that $R_{AV} = 1{\cdot}30$.

4. *Origin at centre of grid*

When the origin is at the centre of the grid the system route factor may be obtained by eliminating the nodes on one or other of the coordinate axes. This reduces both the numerator and denominator by $n(n+1)/2$ grid units. As a result

$$R_N = \frac{n(n+1)(2n+1)}{n(n+1)(1+\sqrt{2})+4\sum_{r=2}^{n}\sum_{s=1}^{r=1}(r^2+s^2)^{\frac{1}{2}}}. \quad (5.6)$$

5. *One-way versus two-way operations*

The comparative route factor for one-way versus two-way operation on a square grid may be obtained by application of the above analysis. The case will be restricted to trips to and from the origin of a single quadrant. It is left to the student to verify the following result:

$$R_C = 1+\frac{3(n+2)}{2(n+1)^2} \qquad \text{when } n \text{ is even,}$$

$$= 1+3\frac{(n+1)^2-4}{2n(n+1)^2} \qquad \text{when } n \text{ is odd.} \quad (5.7)$$

It should be noted that after making adjustment for $n$—odd or even—that a quarter of the outwards trips carry a penalty of 2 grid units. For inwards trips every one suffers a 2-unit penalty and in addition one-quarter of them pay a further 2-unit penalty.

The above result is rather clumsy to extend to the case of a set of random origins and destinations anywhere on a grid. Bunton

F

obtained a solution of this problem by simulation for a $6\times 6$ grid and found the comparative route factor to be 1·48.

The application of the formula given above where only trips to and from the origin are considered gives for $n = 6$

$$R_C = 1{\cdot}25.$$

### *The Ring Radial Network*

Consider a ring radial network with $n$ rings and $m$ equally spaced radials as shown in Fig. 5.2. Consider also that the $n$th circumferential

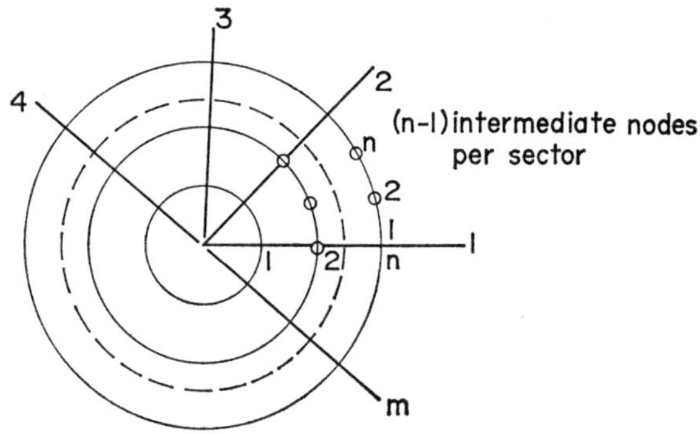


FIG. 5.2. An $n\times m$ ring radial network with intermediate nodes on circumferentials.

has $n$ destinations per sector, i.e. the $n$th ring has $m$ nodes and $m(n-1)$ intermediate destinations. We can thus create a network that has a relatively uniform distribution of destinations. Let the angle between the radials be $= 2\pi/m$. The actual distance* to all points on the $r$th circumferential per sector is

$$r+2\begin{bmatrix} r+\theta \\ r+2\theta \\ \cdot\quad\cdot\quad\cdot\quad\cdot \\ r+\left(\frac{r}{2}-1\right)\theta \end{bmatrix}+\left(r+\frac{r}{2}\theta\right) \qquad \text{if } r \text{ is even}$$

* Assuming unit distance between circumferentials.

and

$$r+2\begin{bmatrix} r+\theta \\ r+2\theta \\ \cdot\ \cdot\ \cdot \\ r+\dfrac{r-1}{2}\theta \end{bmatrix} \qquad \text{if } r \text{ is odd.}$$

The total distance from the origin to all nodes out to and including ring $n$ is thus

$$\tfrac{1}{2}n(n+1)+\sum_{n(\text{odd})}(n-1)\left\{n+\frac{\theta}{4}(n+1)\right\}+\sum_{n(\text{even})}\left\{n(n-1)+\frac{n\theta}{4}\right\}.$$

The total direct distance readily calculated at every point is distance $r$ from the origin, i.e. for each sector

$$1+2{\cdot}2+3{\cdot}3+\dots n^2 = n(n+1)(2n+1)/6.$$

The system route factor is thus

$$R_N = \frac{3}{(2n+1)}+\frac{6}{n(n+1)(2n+1)}\left[\sum_{n(\text{odd})}(n-1)\left\{n+\frac{\theta}{4}(n+1)\right\} + \sum_{n(\text{even})}\left\{n\left(n-1+\frac{\theta n}{4}\right)\right\}\right]. \qquad (5.8)$$

EXAMPLE

What is the system route factor for a ring radial system with $n = 6$ $m = 8$.

$$R_N = \frac{3}{2{\cdot}6+1}+\frac{6}{6(6+1)(2{\cdot}6+1)}\sum_{1,3,5}(n-1)\left\{n+\frac{\theta}{4}(n+1)\right\} + \sum_{2,4,6}\left\{n\left(n-1+\frac{n\theta}{4}\right)\right\}$$

$$= \frac{3}{13}+\frac{1}{7{\cdot}13}\{70+19\theta\},$$

but $\theta = \dfrac{2\pi}{8} = \dfrac{3{\cdot}14}{4}$,

$\therefore \quad R_N = 1{\cdot}165.$

Average distances from centre and circumference to all points within a circle.

From centre $\bar{d} = \frac{2}{3}n$

From circumference $\bar{d} = 32n/9$

where $n$ is the radius of the circle.

Smeed[50] has done notable work in generalising the "kinematics" of land use and transport interactions.

## SHORTEST PATH THROUGH A NETWORK

Although a variety of interesting properties of networks can be developed with the aid of the theoretical results given above, the trigonometrical difficulties that arise in dealing with other than the most simple geometrical networks taken together with the discontinuities introduced by the discrete definition of the nodes limits their usefulness in the general run of practical problems.

However, the development of the theory graph and its application to flows in networks has opened up a powerful new tool for the study of transport systems—the shortest-route and maximum-flow algorithms having a particular relevance. In extending route factor analysis the shortest and *n*th shortest-route algorithms may be applied.

In Chapter 4 algorithms have been described and although they may be readily computerized for application to large-scale problems, it is instructive and useful for the student to acquire some facility in applying them manually to illustrative examples and exercises.

Consider the grid network shown in Fig. 5.3(a). The travel times on the various links are shown. Assuming that one trip originates at every node and is destined for the origin, determine the minimum number of trip miles of travel in the network. Figures 5.3(b) and 5.3(c) indicate respectively a geometrical and operational change in the network. The former introduces a radial route connected at nodes 1, 8 and 12. The latter imposes a one-way routing scheme. Compare the minimum-trip miles in each case with the value obtained in the original network. The shortest path from the origin (node 1) to every other node may be obtained by easy application of the Ford–Fulkerson algorithm. The labels are shown on the respective diagrams. It will be seen that the first modification results in a reduction of 17% in the amount of travel whilst the latter increased by approximately 26%.

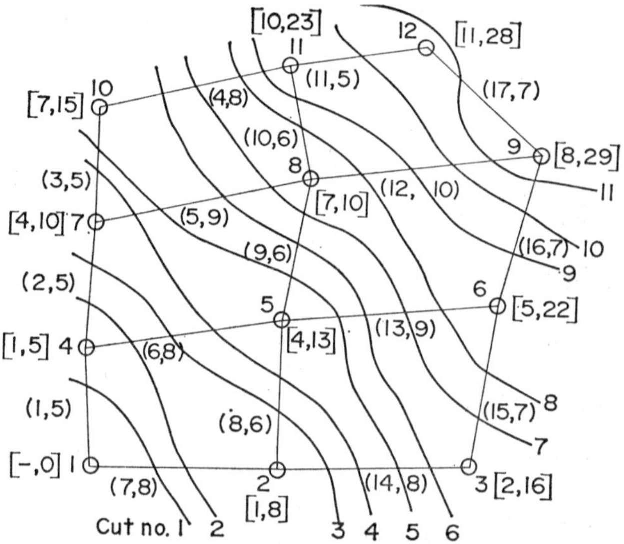


Total Distance from Node 1 to all other Nodes = 188 units

**FIG. 5.3(a).** Shortest paths from node 1 to other nodes for two-way operation.

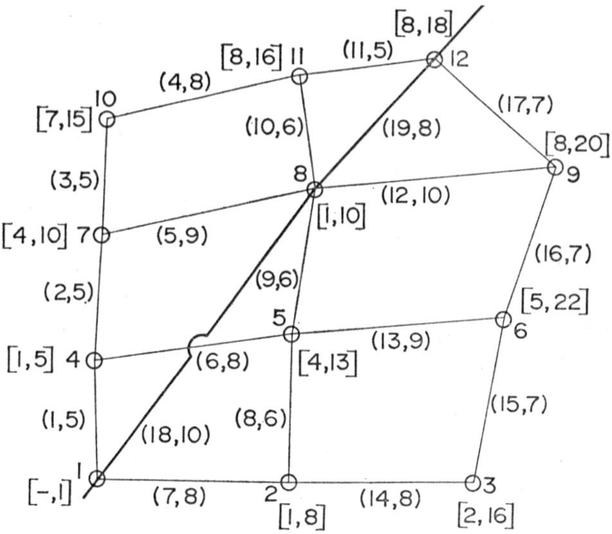


Total Distance from Node 1 to all others = 153 units
Comparative Route Factor = 153/188 = 0·82

**FIG. 5.3(b).** Changes in shortest path as a result of introducing radial links.

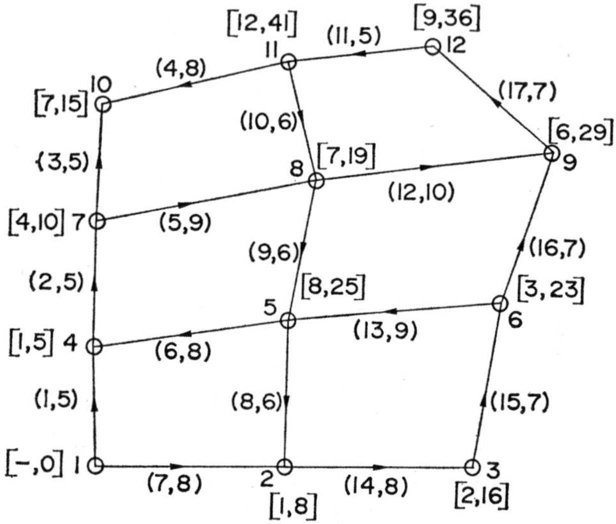


Total Distance from Node 1 to all others = 237 units
Comparative Route Factors 237/188 = 1·26

FIG. 5.3(c). Effect of one-way streets on shortest paths.

### *Less-short Paths*

We have indicated in Chapter 4 two methods of determining the second, third and *n*th shortest paths. The average route-factor or path-length concept based on minimum paths is meaningful as a figure of merit of a network but such a concept based on second or other less short paths between all nodes has little practical value. On the other hand, the ability to determine a limited number of alternative paths is of value in traffic assignment problems. One problem that arises in seeking longer paths is that of looping—in extreme cases traversing a given link in both directions. Another problem is that of determining whether the alternative path should be wholly separate from the shortest or any of the others. The known algorithms do not guard against these possibilities automatically and as a result a good deal of mental surveillance must be exercised by the analyst. The cycling difficulty can in general be eliminated if the network is "directed" and in most practical problems it is appropriate to do this anyway.

Of the two algorithms described in Chapter 4 the Hoffman–Pavley approach is the more useful for finding alternative routes between a given number of *O* and *D* pairs where the number of nodes they include is small compared with the total number in the network. The Bellman–Kalaba algorithm gives a more general solution in that it evaluates all

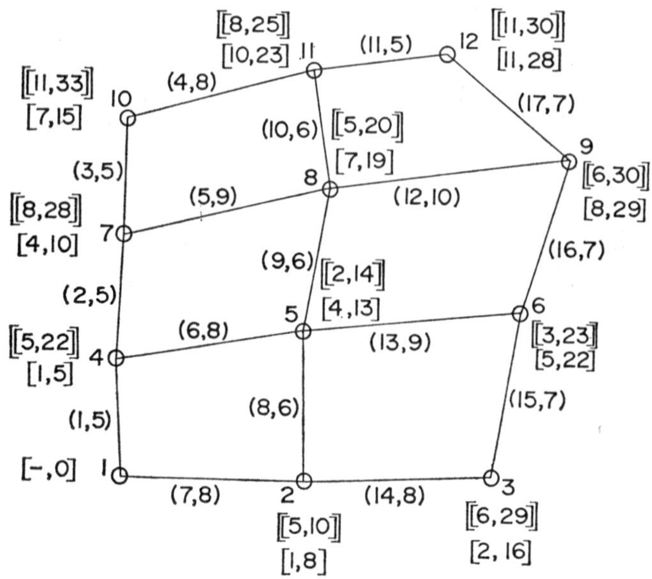


FIG. 5.4. Second shortest paths by Bellman–Kalaba algorithm.

the longer paths from a given origin to every other node. In Fig. 5.4 the Bellman–Kalaba algorithm has been applied to the first network of Fig. 5.3(a) and in 5.3(b) the Hoffman–Pavley method has been used to find the second shortest path between nodes 1 and 12 as shown in Fig. 5.5. The second shortest paths are designated by double-bracketed labels. It is left to the student to check the application of the respective algorithms.

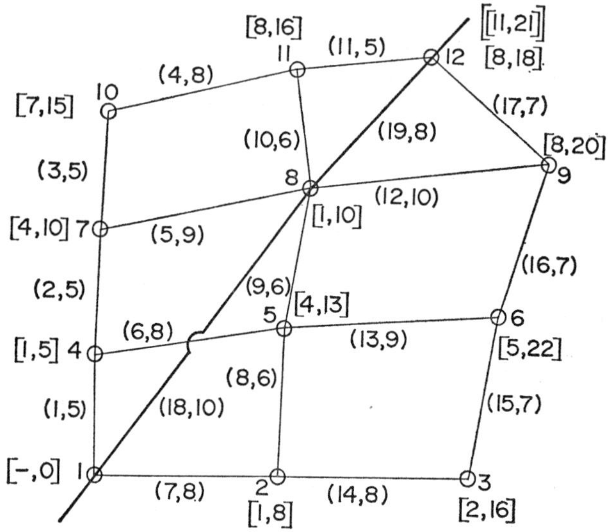


FIG. 5.5. Second shortest path from node 1 to 12 by the Hoffman–Pavley method.[48] *Note*. This is one of the many "deviation" methods. Nodes on the shortest path are examined backwards from the destination. At each such node the path deviates to others joined to it and from each of these the shortest path to the destination is found. The minimum of all such trials is the second shortest path. Some other deviation methods are: (a) remove shortest path altogether and find new shortest path, (b) remove link with minimum capacity from shortest path and find minimum alternative.

### *The Link-path Incidence Matrix*

We have already described the land-use/transport incidence matrix. With less ambitious styling it is also known as the link-path incidence matrix. Generally the incidence matrix may take account of any number of arbitrary paths between *O* and *D* pairs. However, in practical traffic applications it is appropriate to consider reasonable alternative paths which are the shortest and successively less short paths until some specified comparative route factor value is reached.

Consider again the network of Fig. 5.3(a) and consider additionally that desire lines exist from the origin node, 1, to nodes 11, 12 and 6. The network is redrawn in Fig. 5.6. The two element set, the elements represent the serial numbers of the link and its length (or time) respectively.

The link-path incidence matrix may be constructed for the shortest and second shortest paths as shown below

$$l = 1, 2, \ldots, 17, \qquad m = 1, 2 \qquad n = 1, 2, 3$$

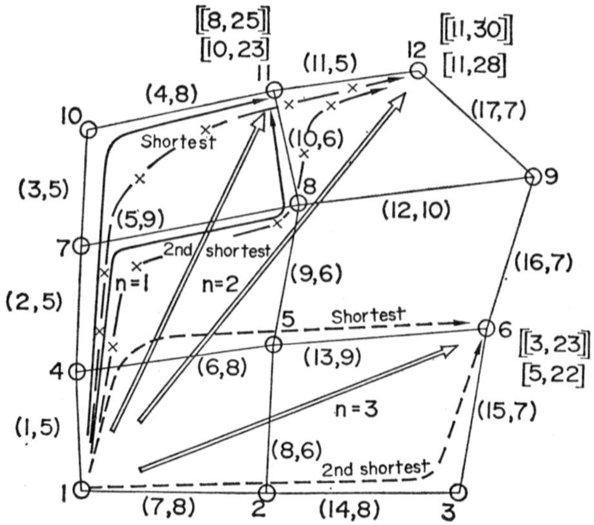


FIG. 5.6. Definition of shortest and second shortest paths for construction of the land-use transport traffic matrix (Table 5.3).

TABLE 5.3. LINK-PATH INCIDENCE MATRIX FOR NETWORK OF FIG. 5.3(a)

| $n$ | $m$ \ $l$ | 1 | 2 | 3 | 4 | 5 | 6 | 7 | 8 | 9 | 10 | 11 | 12 | 13 | 14 | 15 | 16 | 17 |
|---|---|---|---|---|---|---|---|---|---|---|---|---|---|---|---|---|---|---|
| 1 | 1 | 1 | 1 | 1 | 1 | | | | | | | | | | | | | |
| | 2 | 1 | 1 | | | 1 | | | | | 1 | | | | | | | |
| 2 | 1 | 1 | 1 | 1 | 1 | | | | | | | 1 | | | | | | |
| | 2 | 1 | 1 | | | 1 | | | | | 1 | 1 | | | | | | |
| 3 | 1 | 1 | | | | | 1 | | | | | | | 1 | | | | |
| | 2 | | | | | | | 1 | | | | | | | | 1 | 1 | |

SATURATION FLOW—CAPACITY

We have already made reference to the fact that one of the important characteristics of a traffic element is its saturation flow. A transport system comprising a series/parallel arrangement of individual elements exhibits the same property as does a mechanical or electrical system in that the saturation flow for a system of parallel elements is the sum of the individual saturation flows and for a series arrangement it is equal to that of the least capacious element.

The many elements that make up a transport system—e.g. roads, airport runways, railway tracks, bus-loading platforms, ticket windows, baggage check-in points, intersections, ships' berths—fall into two categories, viz.

(a) barrier or cross-sectional,

(b) continuously distributed.

The former group include intersections, baggage channels, berths, runways and an endless variety of other transport facilities. The latter group is for all practical purposes confined to a road lane, a rail track or a pedestrian file. If ships and aircraft were confined to travel in line (and in some circumstances this, in fact, is done), then an air corridor or shipping channel may also be included in the second category.

Traffic-flow elements have a characteristic saturation flow or capacity which is determined by the design features and operational mode of the element. However, there is a need for a sharp distinction to be drawn between barrier and distributed traffic elements so far as their saturation-flow characteristic is concerned. The saturation flow of a distributed element is in general deterministically related to the speed of the flowing objects. Barrier-type elements, on the other hand, are required to come to rest or pause (even if instantaneously) and as a result the saturation flow takes on a single characteristic value—cars pull away from an intersection at about 2-second intervals, aircraft occupy a runway for about 90 seconds, buses load and unload passengers in files at 2-second intervals. Certainly a variety of design and operational details can alter these times and rates, but for established facilities there is a surprising degree of "immutability" about maximum-flow rates of many elemental transport operations. The student might consider again Fig. 3.4(a), (b), (c) and (d).

In the case of continuous elements there is also a maximum-flow rate of a characteristic kind—which could be more convincing than the near-universal maximum-flow rate of 2000 per hour for passenger cars travelling on a level, well-designed road lane. However, in spite of this fact it is important to realize that where independent transport vehicles are constrained to move in line there is an important deterministic relationship between the maximum flow and speed. For example, a line of passenger cars will adjust to a maximum-flow condition of say 2000 vehicles/hour moving between 25–30 m.p.h. On the other hand, if the same line were "caught" behind a heavy truck travelling at, say, 5 m.p.h. on a steep grade then the maximum-flow rate would be of the order of 800–1000 per hour.

Similarly, if the drivers of a military supply convoy were under orders to move at 60 m.p.h. then once again the maximum flow would be much less than the maximum possible flow. The many interesting aspects of the relationship between flow and speed belong more properly to a treatise on the theory of traffic flow. In this section our principal concern will be with the review of saturation-flow rates of the major elements of the different modes and the development of the idea of system capacity.

### *Formal Definition of the Capacity of a Traffic Element*

The saturation-flow rate of a traffic element may be described by the distribution of the time intervals between successive units passing a reference point provided a queue exists upstream of the reference point. The important measures may be described with reference to Fig. 3.2.

$$s(= q_{\max}) = \frac{3600}{\bar{T}_s} \Big/ \text{hour}.$$

In order to take account of the general form of the distribution it is necessary to specify its variance. It has already been made clear (Chapter 3) that the level of service is directly related to the variance of the saturation flow characteristic of a traffic element.

It should be noted that the saturation flow, whether of a barrier or distributed element, can only be defined at a "point" or cross-section.

It is appropriate to stress again that the saturation flow of an element is only meaningfully defined for an individual channel, lane or track. The capacity of any transport complex may be easily synthesized by a series/parallel combination of basic elements.

### *Saturation-flow Values for Various Traffic Elements*

The distributions from which saturation flow parameters are obtained must in general be obtained experimentally. Table 5.4 gives approximate values suitable for concept planning, theoretical synthesis and general evaluations in a wide range of typical situations.

TABLE 5.4.
SATURATION FLOW OF DISTRIBUTED ELEMENTS

| Traffic element | Saturation flow (units/hr) |
|---|---|
| Road lanes | 1800–2200 vehicles |
| Single rail tracks (urban) | 30–35 trains |
| Single rail tracks (main line) | 5–10 trains |
| Exclusive bus lanes (express) | 800–1000 buses |
| Pedestrian lanes | 1600–1800 persons |
| Air lanes | 5–6 aircraft |

Although there are many situations in which it is appropriate to analyse the transport system in terms of the mechanical capacity of its elements in terms of vehicular/units per channel/unit time, there are others in which the principal object relates to the *traffic output*, the basic measures of which are passenger or tons/unit time. The traffic flow is related to the saturation flow of the transport element by the following simple formula:

$$\text{traffic flow} = L_f \times C \times s \tag{5.9}$$

where $C$ is the capacity of the individual transport vehicle in either passengers or tons,

$L_f$ is the load factor.

$L_f$ is a very variable factor and whilst it is generally less than unity there is a notable exception in the case of urban mass transport during peak periods. When one surveys the broad spectrum of transport operations some dramatic comparisons are possible. For example, an

TABLE 5.5. SATURATION FLOW OF BARRIER TRANSPORT ELEMENTS

| Element | Mean service* times | Standard deviation | Saturation flow rate/hr |
|---|---|---|---|
| Intersection | 2·0–2·2 sec | 0·5–0·7 sec | 1600–1800 cars |
| Airport runways (landing or taking off) | 60–120 | 10–20 | 30–60 planes |
| Toll barriers | 5–6 | 2·0 | 600–700 cars |
| Bus stops | 20–30 | 10–20 | 120–180 buses |
| Railway platforms | 60–90 | 20–30 | 40–60 trains |

* These mean service times include the minimum follow-up headway.

TABLE 5.6. CAPACITY OF TRANSPORT VEHICLES

| Vehicle | Capacity | | Ratio of pay load to tare | |
|---|---|---|---|---|
| | Passengers | Freight | Passengers | Freight |
| Aeroplane (jets) | 120–350 | 5–25 tons | 0·2 | 0·4 |
| Train | 100–120/car | 1000–2000 | — | 2–4 |
| Ship | 1000–2000 | 10–50,000 | — | 1 |
| Truck | | 3–20 | — | 2–2·5 |
| Car | 4–5 | | 0·4–0·5 | — |
| Bus | 50–60 | | 0·5 | — |

airport runway used by Jumbo Jets could pass 12,000–15,000 passengers per hour whereas a berth for a passenger liner has a traffic capacity of, say, 200 passengers per hour. The data in Tables 5.4, 5.5 and 5.6 permit interesting comparisons to be made.

### *Maximum Flow Through a Transport System*

It was stated earlier that a transport system may be built up by combining individual transport elements in series and/or parallel. A number of simple arrangements are shown in Fig. 5.7.

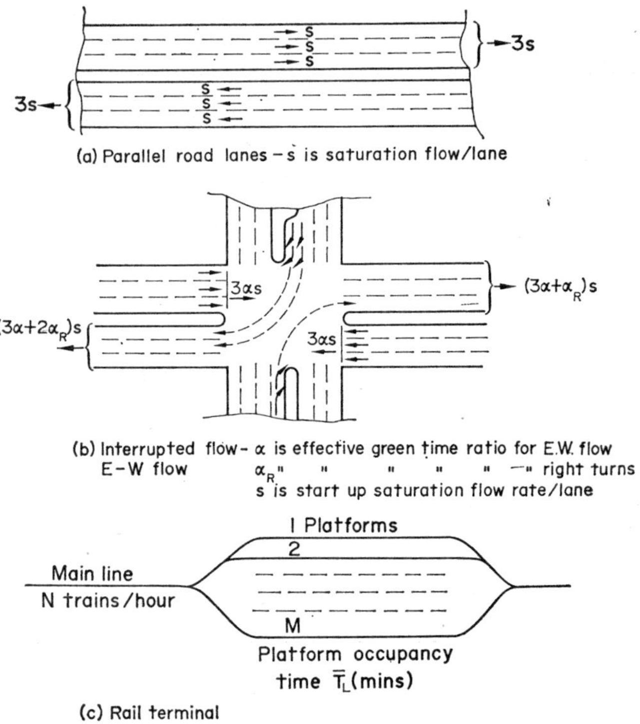


FIG. 5.7. Series/parallel arrangement of basic traffic-flow elements—lanes and tracks.

Consider the situation shown in Fig. 5.7(c).

$$\bar{T}_L = \text{average platform occupancy (minutes)},$$
$$M = \text{number of platforms},$$
$$s_L = 60/\bar{T}_L.$$

For balanced design

$$N = 60M/\bar{T}_L.$$

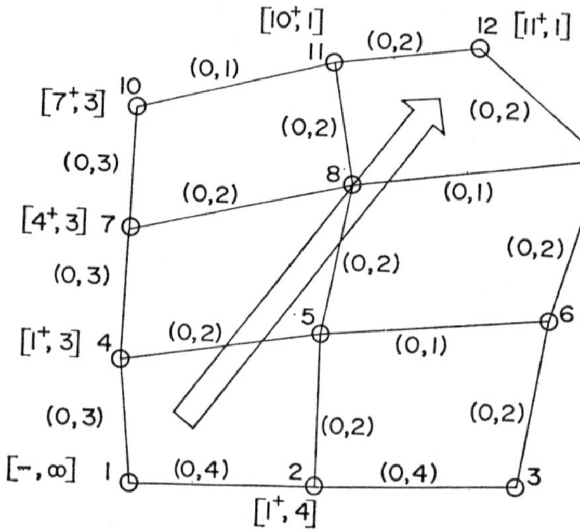


FIG. 5.8(a). Initial labels—assuming zero flows on all links. Number pairs on links give (flow, max. flow) 000's.

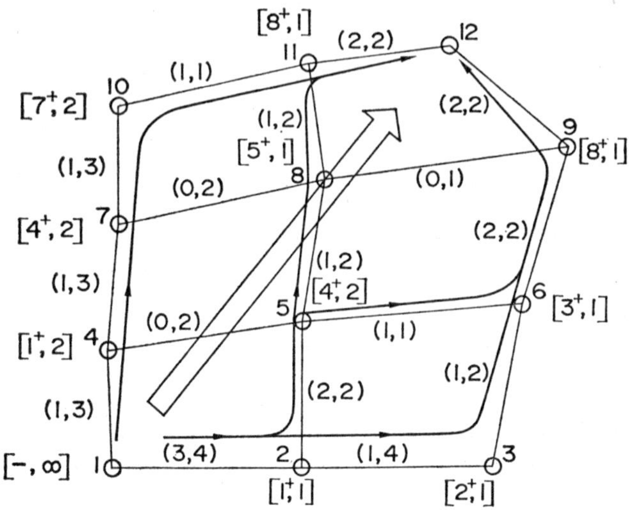


FIG. 5.8(b). Final labelling of the network maximum flow = 4(000's). *Note.* Although max. flow valve is unique there are alternative paths between origin (node 1) and nodes 9 and 11.

If $N < (60M/\overline{T}_L)$ then approach track limits flow if $N > (60M/\overline{T}_L)$ then the terminal provides the bottleneck. In more general terms, for a simple series parallel arrangement of transport elements

$$M_1S_1 = M_2S_2 = \cdots M_nS_n. \tag{5.10}$$

This simple equation is analogous to the equation of continuity for incompressible fluid flow. But in transport system design how often flouted—often at the cost of millions of dollars! On occasions "capacious" new works feed into bottlenecks of a temporary nature, but even with plans for an ultimately balanced system in mind there can be a considerable loss of benefits if consideration is not given to scheduling. In many cases the mismatch of capacity is permanent—particularly at the interfaces between trunk, distributor and local access and terminal components of the system.

Although it is a relatively easy matter to determine by inspection the critical element (elements) in a network of well-defined serial and parallel elements, this is not always the case, particularly in a network with a variety of diagonal links. The maximum flow/minimum cut algorithm explained in Chapter 4 provides the general solution to the problem of finding the bottleneck cross-section (or cut) of any network.

Figure 5.8 establishes the minimum cut and maximum flow for the network described in Fig. 5.3(a) for a desire line between nodes 1 and 12. The network is shown in (a) with its initial labels and the final iteration is given in (b). The reader will have little difficulty in completing the intermediate steps. The critical links are (11, 12) and (9, 12). The maximum flow is 4. As is generally the case in a grid network there are many alternative flow paths between the origin and destination.

## THE EFFECT OF SPEED ON SATURATION FLOW

In the case of flow of vehicles and trains on a distributed element such as length of road or track, there is an important deterministic relationship between maximum flow, speed and density. There are two distinct cases:

Independent manual control,

Automatic or largely automatic control.

*Independent manual control*

This is the situation that applies to all motor vehicle traffic—at least until the automatic highway becomes a reality. It is now well established that the maintenance of a line of independently controlled vehicles in a state of stable motion depends on the driver of each vehicle carrying out a velocity (or first derivative) tracking task with respect to the vehicle in front of him. In more specific terms the acceleration (or deceleration) that a driver, following a vehicle in front of him, gives to his vehicle is proportional to the relative velocity between his and the vehicle ahead, viz.

$$\frac{dV_2}{dt} = K(V_1 - V_2).$$

However, there are two refinements of this basic process that are essential in describing a real situation—the second vehicle's response is delayed by the response time of a driver–vehicle unit and the sensitivity constant, $K$, is inversely proportional to the spacing. The saturation flow equation may thus be written

$$\left[\frac{dV_2}{dt}\right]_{t+R} = \frac{K}{x}[V_1 - V_2]_t$$

where $R$ is the driver–vehicle response lag.

It is not intended to pursue this interesting equation except to give its steady-state solution

$$d = d_0 e^{-V/V_c} \tag{5.11}$$

where $d$ is the density (i.e. $1/\bar{x}_s$), where $\bar{x}_s$ is average distance between cars when the stream has settled down in uniform motion at speed $V$,

$d_o$ is the maximum density, i.e. the density when the stream has just come to rest,

$V_c$ is the speed for maximum flow.

The saturation flow for a given speed is

$$S(= q_{max}) = V \, . \, d_o e^{-V/V_c} \tag{5.12}$$

and the "maximum" saturation flow at the optimum speed is

$$S_{max} = 0{\cdot}37 V_c d_o.$$

For example, if $d_o = 200$ vehicles/mile,

$V_c = 25$ m.p.h.,

then $S_{max} = 1850$ vehicles/hour.

If now a slow-moving vehicle (perhaps a heavy truck on a grade) moving at 5 m.p.h. were to create a long queue of vehicles behind it, then the saturation flow would be

$$S = 5{\cdot}0 \times 200e^{-5/25}$$
$$= 820 \text{ vehicles/hour.}$$

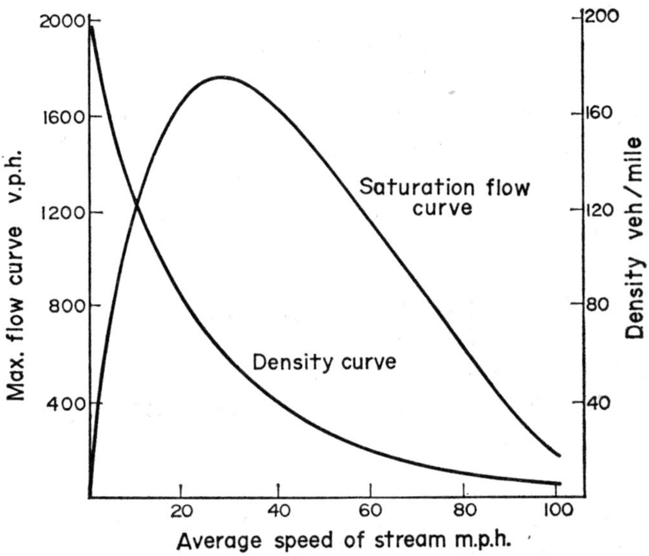


FIG. 5.9. Saturation/speed and density/speed curves for steady flow of line of motor vehicles.

Alternatively, a military convoy of staff cars may be ordered to travel at 50 m.p.h., then

$$S = 50 \times 200e^{-50/25}$$
$$= 1400 \text{ vehicles/hour.}$$

Figure 5.9 shows how the flow and density of road traffic varies with the average speed of the stream. It should be noted that the average speed is determined from a space (longitudinal) distribution of speeds at a given instant in time.

Perhaps one should conclude this brief discussion by noting that the reciprocal of independent variable (i.e. the speed) must not be confused with the travel time or delay. It is essentially a service rate or apparently paradoxically a "zero-flow" travel time.

*Non-manual control*

In the case of rail and air traffic and in certain other transport situations the role of the individual driver or pilot is heavily supervised by automatic or regulatory control. In general control is based on the principle of ensuring that a safe follow-up distance or headway is maintained between successive vehicles. A safe distance, $x$, front to front of vehicles travelling at a speed $V$ is clearly

$$x = L+RV+BV^2$$

where $L$ = length of vehicle,

$R$ = response time of detection and reaction to change of motion of a lead vehicle,

$B$ = the braking coefficient;

thus

$$d = \frac{5280}{L+RV+BV^2} \quad (5.13)$$

and

$$q_{\max} = \frac{5280V}{L+RV+BV^2}. \quad (5.14)$$

It should be noted that whilst it is not theoretically necessary for the denominator to include the braking term if it can be assumed that each unit has equal braking capability the "fail-safe" practice which applies in all common carrying operations demands that the safe distance should not only be based on the full stopping distance but should include some additional factor of safety.

TRAVEL TIME

We have already considered the fundamental relationship between travel time and traffic intensity which applies in virtually all transport situations. The total travel time is the sum of a constant (or near constant) component already defined as the zero flow travel time and a variable component which depends on the traffic intensity and which increases very rapidly as the flow approaches the saturation flow.

Our concern here is with the former component which is primarily a function of the physical characteristics of the transport apparatus—acceleration, maximum speed of the transport vehicle, grades and curves of the highway or track. One might include also certain operational characteristics such as frequency or service, number of stops of mass transport systems. It is of some importance to state that zero flow travel times are susceptible to reasonably accurate theoretical determination from the dynamical equations of motion. They are also susceptible to reasonably straightforward field measurement.

### *The Equation of Motion*

The equation of motion of any transport vehicle may be written:

$$M\frac{d^2x}{dt^2} = f\left(\frac{dx}{dt}\right) \pm g(x) \qquad (5.15)$$

where $x$ is the distance travelled from some suitable origin,

$f\left(\frac{dx}{dt}\right)$ is the tractive force developed at the driving wheels, propellors or jets less air resistance and drag and dynamic braking forces,

$g(x)$ is external loading on the vehicle—for road and rail vehicles it is made up of the grade and curve resistance and applied braking forces; in the case of aeroplanes in climbing motion it is the force of gravity,

$M$ is the mass of the vehicle.

In the case of sea and air travel the determination of journey time must of course take account of currents and winds.

This equation has been much exploited in determining railway timetables and schedules. $f(dx/dt)$ is in general an experimentally determined function and the solution of the equation is usually obtained by numerical integration. However, the tractive force function has certain quite definite characteristics that are well defined. Figure 5.10 indicates

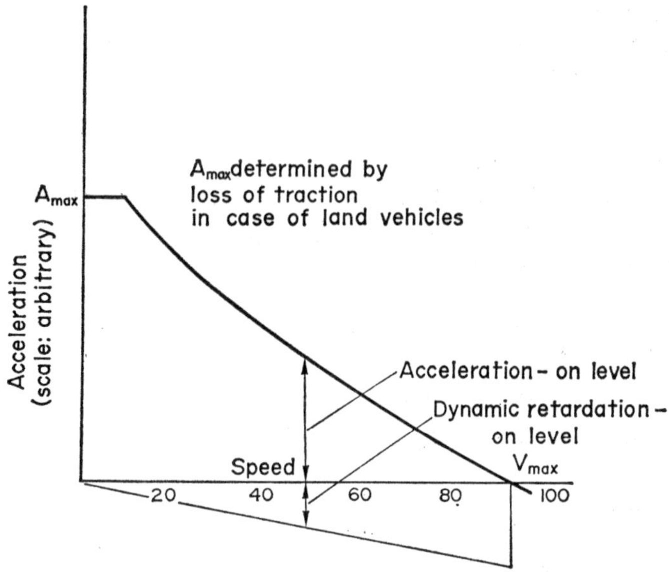


FIG. 5.10. Acceleration and dynamic braking characteristics of prime movers—typical only.

its general form. If the equation is divided through by $M$ the various terms become accelerations or retardations and integration leads directly to speed/distance and time/distance curves.

The dynamic braking characteristic varies from a maximum value at top speed to zero at rest. In the case of motor cars the magnitude of the dynamic deceleration is small compared with the decelerating power of the brakes and in any case the driver has peculiar ability to adjust his brake power, give a steady deceleration of magnitude appropriate to the circumstances. Typical values are

| | |
|---|---|
| gradual | 2–4 ft/sec$^2$ |
| normal | 4–6 ,, |
| emergency | 16–24 ,, |

Apart from the application of the equation of motion to the problem of programming and scheduling, it has application in the road traffic context to the determination of vehicle trajectories in overtaking and other common traffic manoeuvres.

### *Transport Vehicle Trajectories*

If the acceleration/speed curve is represented by a straight line then its slope $C$ is negative and equal to $A_{max}/V_{max}$. The equation of motion on a level straight is thus

$$\frac{d^2x}{dt^2} = \frac{dV}{dt} = A_{max} - \frac{A_{max}}{V_{max}} . V.$$

$$= A_{max} - CV \qquad \text{accelerating.}$$

For modern cars $A_{max} = 12–15$ ft/sec$^2$; $C = 0{\cdot}10–0{\cdot}12$.

$$\frac{d^2x}{dt^2} = \frac{dV}{dt} = 0 \qquad \text{running at constant speed,}$$

$$\frac{dV}{dt} = -B \qquad \text{braking.}$$

These equations may be used to make realistic calculations of

(a) times to drive into or cross a gap,
(b) overtaking times and distances,
(c) travel time between intersections,
(d) time tables and schedules.

Even the simple linear approximation to the acceleration characteristic of a transport vehicle causes a great deal of algebraic difficulty in solving practical problems. However, the following results are worth stating.

They apply to a vehicle accelerating from rest on a level roadway or track.

(i) Time to reach a speed $V$

$$t = \frac{1}{C}\log\frac{A_{max}}{A_{max}-CV}. \tag{5.16a}$$

(ii) Speed after time $t$

$$V = V_{max}\left[1-e^{-Ct}\right]. \tag{5.16b}$$

(iii) Distance travelled after time $t$

$$x = V_{max}\left[t+\frac{1}{C}(e^{-Ct}-1)\right]. \tag{5.16c}$$

(iv) Distance travelled when speed has reached $V$

$$x = \frac{V_{max}}{C}\left[\log\frac{A_{max}}{A_{max}-CV}-\frac{C}{A_{max}}V\right]. \tag{5.16d}$$

*Minimum travel time between stops*

This is a problem of some considerable general interest in transport scheduling. In order to simplify the analytical treatment we will assume that the vehicle has constant acceleration $A$ until it reaches a running speed $V_R$ and that it brakes with a constant deceleration $B$. If the separation of the two stops is $D$ then the journey time is

$$t = V_R\left(\frac{1}{A}+\frac{1}{B}\right)+\frac{D}{V_R}.$$

The running speed for minimum journey time is

$$V_{min} = 2DAB/\sqrt{(A+B)} \tag{5.17}$$

provided

$$V_{min} < V_R.$$

It will be instructive for the student to ponder the significance of the sensitivity of $V_R$ to the values of $A$ and $B$. If $D \doteq \frac{1}{4}$ mile and $A = 5$ ft/sec$^2$ and $B = 10$ ft/sec$^2$ then $V_R = 62$ m.p.h. approx. On the other hand, with $A = 2$ and $B = 5$, $V_R$ is just 30 m.p.h.

*The line haul passenger service*

In a research project at the School of Traffic Engineering, Hughes has obtained the following result:[51] the total time spent by passengers in transit is

$$t = \left[\frac{D}{V_R}+\frac{V_R}{2}\left(\frac{1}{A}+\frac{1}{B}\right)\right]\frac{NM(1+M)}{2}+\frac{N^2M(M-1)}{2q}+\frac{NM}{2f} \qquad (5.18)$$

where $D$ $V_R$, $A$ and $B$ are as defined above;

- $N$ is the number loading at each stop—assumed same for each stop;
- $q$ is the rate of loading;
- $M$ is the number of stops—all distance $D$ apart;
- $f$ is the frequency of service.

In the case of buses operating in the general traffic stream, theoretical calculations are in general inappropriate and regression techniques applied to real traffic data prove more practical. Smeed has presented data on bus travel times in Central London and produced a very useful relationship between travel time of buses and that of the general traffic stream, viz.

$$\text{Bus time} = (\text{car time} + 1{\cdot}5) \text{ minutes/mile.} \qquad (5.19)$$

*Bus travel time in* ANZAC *City*

In *Introduction to Traffic Science*[2] the author proposed a bus transport system similar to that shown in Fig. 1.5. The operational data pertaining to this system may be summarized

| | |
|---|---|
| Frequency of service, N.–S. routes: | 6 per hour |
| Frequency of service, E.–W. routes: | 3 per hour |
| Stops at all nodes of the ½-mile grid | |
| Walking speed | 3 m.p.h. |
| Average concentration and dispersal distance about zone centroid | ¼ mile. |

TABLE 5.7. BUS TRAVEL TIMES FOR ANZAC CITY

| Destination | Mode | Service frequency N.–S. (per hour) | Service frequency E.–W. (per hour) | Origins 2 | 3 | 4 | 5 | 7 | 9 | 10 | 11 | 12 |
|---|---|---|---|---|---|---|---|---|---|---|---|---|
| 1 | Car | — | — | 6·3 | 4·5 | 11·5 | 6·5 | 17·0 | 12·5 | 14·5 | 16·0 | 14·0 |
| | Bus | 6 | 3 | 21·8 | 15·8 | 33·5 | 19·3 | 50·3 | 38·3 | 41·8 | 32·5 | 30·5 |
| | | 12 | 10 | 19·3 | 13·3 | 31·0 | 17·8 | 47·8 | 29·3 | 32·8 | 30·0 | 28·0 |
| | | 24 | 20 | 18·0 | 12·0 | 29·8 | 16·5 | 46·5 | 26·5 | 30·8 | 28·8 | 26·8 |
| 6 | Car | — | — | 13·3 | 9·5 | 13·5 | 7·5 | 18·5 | 7·0 | 7·0 | 17·5 | 14·5 |
| | Bus | 6 | 3 | 40·8 | 30·3 | 38·5 | 21·8 | 68·3 | 21·3 | 29·5 | 42·5 | 38·0 |
| | | 12 | 10 | 31·3 | 20·8 | 31·5 | 14·8 | 61·3 | 14·3 | 22·5 | 33·0 | 29·5 |
| | | 24 | 20 | 28·5 | 18·0 | 30·0 | 13·3 | 59·8 | 12·8 | 21·0 | 30·3 | 25·8 |
| 8 | Car | — | — | 13·8 | 10·0 | 11·0 | 8·0 | 13·5 | 8·0 | 10·0 | 12·5 | 9·5 |
| | Bus | 6 | 3 | 38·5 | 18·0 | 31·8 | 14·5 | 43·5 | 19·5 | 23·0 | 31·3 | 16·8 |
| | | 12 | 10 | 29·0 | 15·5 | 24·8 | 12·0 | 36·5 | 12·5 | 16·0 | 21·8 | 14·3 |
| | | 24 | 20 | 26·3 | 14·3 | 23·3 | 10·8 | 35·0 | 11·0 | 14·5 | 19·0 | 13·0 |

*The effect of land-use density on public transport travel times*

| | | |
|---|---|---|
| High density to high density −3·0 | MD to HD −1·5 | LD to HD +1·0 |
| Medium density −1·5 | MD as in table | MD +2·5 |
| Low density +1·0 | LD +2·5 | LD +5·0 |

The total journey time is the sum of

concentration time,

walking time from zone centroid to bus stop,

waiting time (including waiting time for a change of bus),

travel time of bus—obtained from Smeed's formula.

The travel times between all nodes are given in ref. 2, table VII, p. 32. Smith(52) has extended this illustrative example by considering a number of different bus networks and different land-use densities, i.e. low, medium and high having densities of 10,000, 20,000 and 50,000 persons/square mile respectively. This has the effect of reducing the concentration and dispersal distances. Smith also considered bus-assignment ratios approximately double and quadruple that taken by the author and as a result the waiting times were reduced inversely as these ratios. A summary of his results for a bus system in which the two N.–S. to E.–W. through routes are replaced by N.–S. and E.–W. loops is given in Table 5.7. Only trips between residential zones and work zones are shown as the analysis is appropriate only to work trips. Car travel times are shown for comparative purposes.

## TRANSPORT COSTS

The student is confronted with many difficulties in attempting to come to grips with the problem of transport costs. Even in the case of road transport where a considerable amount of good data is available on the direct operating costs to the user there is no agreed on basis for the apportionment of the road costs amongst the various classes of user. When it comes to the main mass modes, operating cost data has an extremely confidential aspect—partly for commercial reasons and partly because of the complex system of subsidies and cross subsidies over the industry as a whole or within particular parts of it.

It will be necessary therefore to base the discussion of this topic on costs to the user and in the case of the mass modes the cost to the user will be taken in the main to be the fare or the freight rate.

*Some basic ideas*

Although this section is styled transport costs it is of fundamental importance to appreciate that the cost of a trip is made up of a number of components:

(a) for freight—transport, inventory, insurance and packaging;
(b) for passenger trips—out-of-pocket costs (i.e. fares, running expenses), time, comfort and so on.

Generally the various components of the trip cost may be classified into three components:

fixed costs—depreciation due to age, garaging, interest, wages, etc.;

running costs—petrol, tyres, maintenance, depreciation due to mileage;

time costs related to the value of the trip—inventory, insurance, packaging, private time.

These three components when summed make up the cost of distribution or the cost of the journey.

***Freight and Goods Transport***

The distribution cost for the transport of goods may be written

$$D = T+S+P+I \tag{5.20}$$

where $T$, $S$, $P$ and $I$ are respectively the transport, storage, packaging and insurance components of the total distribution cost. We shall consider the various components of this equation.

The transport cost $T$ may be written

$$T = D \,.\, K_R \,.\, K_T \,.\, C \,.\, Q \,.\, K_S \tag{5.21a}$$

where $D$ is the direct or airline distance,

$K_R$ is the route factor,

$K_T$ is the tare factor which takes account of the effectiveness of space utilization of the vehicle,

$K_S$ is the profit or subsidization factor or arbitrary additions or subtractions of funds to or from the transport account by way of subsidies, concessions or unrealistic joint costs,

$C$ is the true average operating cost in say cents/ton mile,

$Q$ is the quantity of goods in say tons.

Alternatively the transport cost may be expressed

$$T = D \,.\, K_R \,.\, Q \,.\, R \tag{5.21b}$$

where $R$ is the freight rate, or

$$T = FC + VC + P \tag{5.21c}$$

where $FC$ is the sum of all fixed costs per trip, i.e. stations, rolling stock, vehicles, control equipment,

$VC$ is the sum of variable costs per trip, i.e. fuel, power, maintenance and management,

$P$ is the profit/trip.

Equation (5.21a) is of particular value for analysing and comparing various modes with each other. However, in the general area of transport costing there is such a paucity of reliable data on factors such as $K_s$ and $C$ that this form of the equation is of little use for estimating the cost of shipping. Equation (5.21b) is useful to the shipper and equation (5.21c) is the meaningful formula to the transport operator.

### *Freight Rates and Operating Costs*

Tables 5.8 and 5.9 give typical freight charges for rail and road transport in Australia. Freight rates vary considerably from country to country. Even so there is a good deal of "internationalism" in the area of transport—at least in countries with a comparable standard of living. The main reason for including the above tables, however, is to provide the student with some data for exercises and assignments. The tables also illustrate one particularly important fact concerning freight transport operations—the dominating influence of terminal operations on costs—this being inferred from the rapid increase in unit cost with decreasing haul.

TABLE 5.8. TYPICAL RAIL-TRANSPORT FREIGHT RATES(49)

| Type of freight | Distance (miles) | Freight rate (c/ton-mile) |
|---|---|---|
| Wool (greasy) | 100 | 13·0 |
| Seven bales/ton | 200 | 10·5 |
| | 400 | 7·5 |
| Cement, woolpacks, etc. | 100 | 8·0 |
| | 200 | 6·0 |
| | 400 | 4·5 |
| Iron and steel sections | 100 | 8·5 |
| | 200 | 6·5 |
| | 400 | 5·0 |
| Machinery, foodstuffs, etc. | 100 | 12·5 |
| | 200 | 11·0 |
| | 400 | 8·5 |

TABLE 5.9. TYPICAL ROAD-TRANSPORT FREIGHT RATES(49)

| Type of freight | Distance (miles) | Freight rate (c/ton-mile) |
|---|---|---|
| Terminal distribution | | |
| Wool (7 bales/ton) | 5 | 35 |
| | 10 | 28 |
| | 20 | 19·5 |
| Iron and steel sections | 5 | 41 |
| | 10 | 29 |
| | 20 | 20 |
| Machinery | 5 | 67 |
| | 10 | 40·5 |
| | 20 | 26 |
| General long distance | 50 | 8·0 |
| | 100 | 6·5 |
| | 200 | 5·5 |
| | 400 | 5·0 |

This is a salutary point for the conventional traffic engineer and analyst. How often is he preoccupied with savings of 0·5 cents per mile or saving a minute of travel time forgetting altogether that a possible consequence of his new freeway will be an extra 10 minutes cruising for a parking space.

TABLE 5.10. TYPICAL OPERATING COSTS AND FREIGHT RATES FOR LONG-DISTANCE TRANSPORT IN AUSTRALIA AND THE UNITED STATES[49]
(U.S. data converted at parity)

| Mode | Australia (cents/ton-mile) | | U.S.A. (cents/ton-mile) | |
|---|---|---|---|---|
| | Operating | Freight rate | Operating | Freight rate |
| Rail | 2·5–6 | 4–10 | 1·25–3 | 2–5 |
| Road | 3–6 | 5–8 | 3–6 | 5–8 |
| Air | 20–35 | 40–60 | 10–18 | 24–32 |
| Sea (bulk cargo) | 0·2–2·0 | 0·25–2·5 | 0·2–2·0 | 0·25–2·5 |
| Pipeline | 0·1–1 | — | 0·1–1 | — |

TABLE 5.11. THE RELATIVE MAGNITUDES OF THE FIXED AND VARIABLE COSTS IN FREIGHT TRANSPORT[49]

| Mode | % Distribution | |
|---|---|---|
| | Fixed | Variable |
| Rail | 60 | 40 |
| Road—Long distance | 15 | 85 |
| Local | 30 | 70 |
| Sea | 50 | 50 |
| Air | 85 | 15 |

Table 5.10 gives some rough idea of the ratio of operating (excluding fixed) costs to freight charges in Australia and the U.S.A. Table 5.11 indicates the order of the relative magnitudes of fixed and variable costs for the main transport modes.

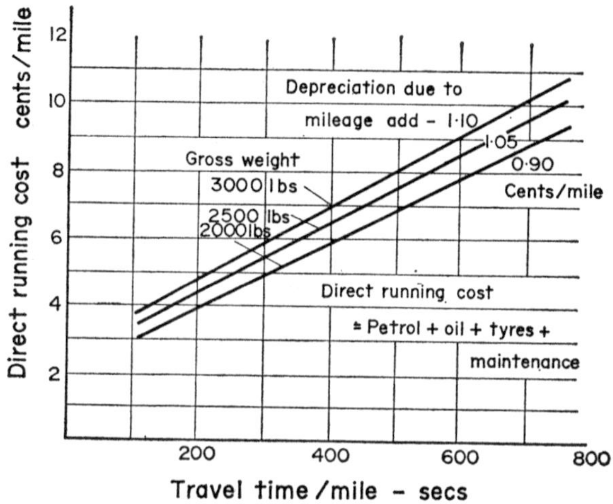


FIG. 5.11. Direct running costs of cars in urban traffic.(54)

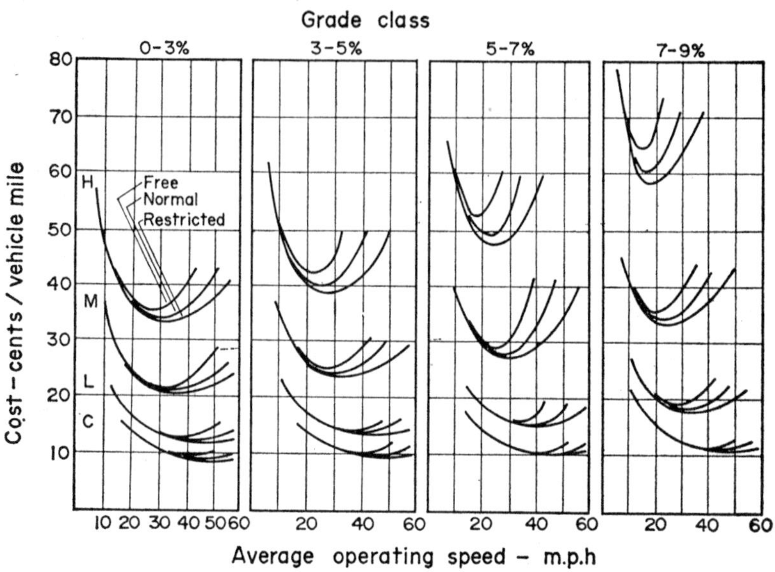


FIG. 5.12. Typical vehicle-operating costs of different vehicle types, on different grades as a function of average operating speed.(53)

One notable exception to the paucity of data on the true operating costs of transport relates to the direct operating costs of road vehicles. Two notable studies carried out by the late Eugen Pelensky[53,54] give authentic data on the unit operating costs of cars and trucks on rural roads and in urban traffic. Figures 5.11 and 5.12 indicate the basically different nature of the cost characteristics for rural and urban operations. In the former grades, operating speed, alignment, pavement quality are all important, in the latter total travel time is the dominant variable including as it does intersection delays, pedestrian interaction, traffic manoeuvres of all kinds. The urban cost characteristic taken together with the travel time/flow relationship is of importance in all urban-planning and economic analysis.

### *Storage, Packaging and Insurance—$S+P+I$*

Although the transport cost is the more readily identifiable component of the distribution cost, the "external" cost factors storage, packaging and insurance have a major role to play in transport operations. It is now well known that direct out-of-pocket costs alone do not explain passenger transport behaviour. Similarly preferences for and assignment to the different freight modes are not determined by the direct operating cost. Storage, packaging and insurance costs are proportional to the product of the value of the freight and the time in transit. It is not too difficult to classify freight in terms of its dollar value per unit weight. Figure 5.13 indicates the general nature of the distribution cost as a function of the value of the freight. Such curves provide a rational basis for freight transport assignment. Table 5.12 gives preliminary basic data for proximate calculations.

For passenger transport the external time cost may be taken as the travellers wage rate or some more or less arbitrarily determined proportion of it. An average acceptable figure currently used in Australian transport planning exercises is \$1·0/vehicle hour.

## THE TRANSFER TERMINAL

The term transfer terminal is used here because the facilities usually known as terminals, i.e. docks, airports, railway stations, are in truth

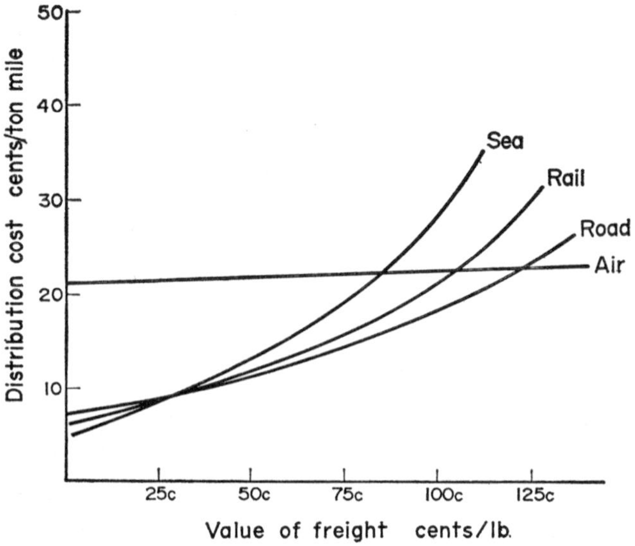


FIG. 5.13. Total distribution cost $(T+S+P+I)$ for various modes.

TABLE 5.12. STORAGE, PACKAGING AND INSURANCE COSTS

| Value of freight | Low value (up to 10c/lb) | High value, ($/lb) |
|---|---|---|
| Storage | 3 cents/ton/day | 100 cents/ton/day |
| Packaging | 0 | 2 cents/lb |
| Insurance | 0 | 0·01 cents/lb/mile |

mode transfer devices. And if walking is recognized as a mode then parking lots, bus stops, lift plazas fall within the definition of the transfer terminal.

Transfer terminals may be represented by miniature transport networks consisting of a series/parallel arrangement of traffic elements. In general the traffic elements are of the barrier type. For example, a kerbside bus stop is a simple single-channel queueing element. A parking lot is a multichannel queueing facility. Baggage check-in counters at airports, departure lounges, plane-loading operations, queueing for take-off clearances all represent a sequence of queueing elements. This sequence of barrier elements is in general linked together by a series of distributed traffic elements utilizing the walking mode.

There are a number of distinct types of problems that confront the analyst of a terminal network, viz.

(a) conventional flow problems,
(b) critical path flow problems,
(c) inventory or storage problems.

Generally, an essentially similar form of analysis needs to be carried out separately for the movement of passengers and freight through a terminal. In the latter situation particular attention must be given to the provision of storage space for freight items. Storage space is required for passengers in the form of queueing space behind ticket windows and baggage counters, in transit lounges and so on. Although important it is an aspect that is usually taken care of in the architectural-design phase. In the case of freight, however, there is need for more serious calculations to be made concerning the amount of warehouse and open-air storage space. There is need too to consider the space requirements for the transport vehicles, i.e. bus bays, railway platforms, ship berths and tarmac space for aircraft. In addition terminal facilities are characterized by the need to provide a holding or reservoir space to accommodate the queue. This kind of storage has a variety of forms—the holding pattern over an airport, the harbour anchorages or the reservoir areas or public street space for motor vehicles.

Although there is great scope for much analysis of terminal traffic situations, little formal development of this important subject has as yet taken place. A most ambitious study by Rallis(18) on the *Capacity of Transport Centres—Ports, Railway Stations, Road Haulage Centres and Airports* should provide a good deal of stimulation for further work. Other recent studies by Jones and Blunden,(55) Helly(56) and Hill(57) are of interest. The first deals with ship-turn-around times in the Port of Bangkok, the second with cruising for parking spaces and the third with the distribution of the amount of cargo on the wharf in transshipment from ship to truck convoys.

*Analytical Methods*

Queueing Theory and Mathematical Programming provide a comprehensive methodology for the evaluation of operational behaviour and for the planning of new facilities. A good deal of basic data needs to be assembled to aid the application of theoretical analysis. In particular

(a) service-time distributions of the many and varied activities of terminals—parking times, check-in services, ticketing, truck loading and unloading times, berth occupancy and so on;
(b) cost functions for service activities—wage rates, capital charges for service facilities;
(c) value of time of passengers, inventory and storage costs, demurrage charges.

The criteria for evaluation are

(a) probability of rejection,
(b) probability of delay,
(c) average delay and average queue length,
(d) probability of delay or queue length exceeding specified operational standards,
(e) minimum total operational cost.

EXAMPLES

Two examples have already been worked out in Chapter 3. One deals with the delays to buses at a kerbside bus stop and the other concerns a five-channel loading dock. Let us pursue the loading-dock problem. In the Example quoted the average delay in the queue was 7·7 minutes. If now the traffic load increases from 3·5 to 4·5, i.e. the arrival rate increases to 7·8 vans/hour, then a reworking of the problem shows that the delay in the queue increases from 7·7 minutes to 55 minutes and on 5% of occasions there would be in excess of 24 vehicles waiting to use the docking facility. This is quite a salutary result. Simple calculations of this kind provide the operations manager or the planner with much-needed clues for their decisions especially in situations such as this where the behaviour of the system is characterized by such a high degree of non-linearity.

It is left to the student to assess the effects of introducing a new loading dock thus increasing $M$ from 5 to 6.

### *The Parking Problem*

The application of certain parallel-channel queueing formulae provides a good deal of enlightenment on the operation of parking facilities of all kinds.

The essential interpretation of queueing parameters in this context is

(a) the service time is the parking duration and the mean service time the average parking time;

(b) the variance of the parking-time distribution needs to be checked experimentally but in the absence of data it is reasonable to assume that parking-time distributions are exponentially distributed;

(c) the number of channels is equal to the number of parking bays.

The principal measure of performance of a parking lot is the Probability of Rejection, $P_L$.

If one consults Table 5.13 it is clear that $P_L$ decreases rapidly as the number of parking bays increases even when the traffic load approaches $M$.

An important implication of the Table 5.13 result is that provided restrictions to the inbound flow do not occur in the aisles near the entry points there is little need to consider reservoirs for large car parks. On the other hand, in the case of small groups of kerbside parking and loading zones $P_L$ is appreciable and as a result considerable "cruising" may occur. The above treatment assumes, of course, that the demand does not exceed the total service rate. Whilst it is not possible to guarantee that $A$ is always less than or equal to $M$ the extreme non-linearity of the delay/flow characteristic will exert a powerful influence in ensuring that the demand is effectively controlled.

Where entry formalities, i.e. boom gates, ticketing booths, are employed, reservoir space may be essential. Data will, of course, be required on the service-time distributions of the entry facility. Boom gates have mean service times between 5–10 seconds. Inbound ticketing service times are of the same order. Outbound pay booths have service times approximately double those of entry facilities.

TABLE 5.13.

| Number of bays, $M$ | $P_L$%*($A \to M$) |
|---|---|
| 1 | 50 |
| 10 | 21 |
| 50 | 14 |
| 100 | 8 |
| 1000 | $\to 0$ |

* It is of relevance to note that the Erlang loss formula is valid for an arbitrary service-time distribution.

## PERFORMANCE MEASURES

We have already referred briefly, in the Introductory Chapter, to the question of the overall performance of a land-use/transport system and we shall take this matter up again in Chapter 10. At this point we will restrict our consideration to the behaviour of the transport system only. But even in this more limited context it is not clear whether induced (or generated) traffic rightly belongs to the lesser or the wider context. For convenience we shall, more or less arbitrarily, exclude it here and return to it in Chapter 10. Our problem then becomes one of assessing the efficiency of the transport system under a specified traffic load.

Two broad approaches may be taken—one based on economic performance and the other employing physical measures of behaviour. In the section dealing with operating costs we have already referred to fixed and variable operating costs of the vehicle. However, consideration of the overall economic merit of transport systems, including the associated investment in the permanent facilities, i.e. roads, rail tracks, ports, is a matter more appropriate to the evaluation phase. We will thus concentrate here on the physical assessment of performance.

***Transport System Output***

The output of a transport system may be considered in terms of

(a) trips, i.e. passengers or tons,

(b) passenger miles (hours), ton miles (hours).

The first is analogous to a vector and the latter a scalar quantity. The latter is useful therefore in defining the output of a system comprising a variety of separate transport elements as the passenger or ton miles of transport activity on the individual components are simply additive.

It is necessary to introduce a time specific measure in order to effect meaningful comparisons, i.e. trips/unit time, ton miles/unit time or passenger hours/unit time, and we may describe such measures as the specific outputs.

Consider a length of road (say) of length $l$, then the flow of vehicles along it may be written

$$q = v d_o e^{-v/v_c}.$$

Differentiation with respect to $v$ shows that $q$ is maximum when $v = v_c$ and

$$q_{\max} = v_c d_0/e.$$

If now we take account of the transport task then the vehicle miles/unit time may be written

$$U = q\frac{l}{t} = qv = v^2 d_o e^{-v/v_c}.$$

Differentiation with respect to $v$ shows that for maximum output

$$v = 2v_c$$

and

$$U_{\max} = 4v_c^2 d_o/e^2.$$

Figure 5.14 shows the specific output function plotted against speed for a road link with $d_o = 200$ vehicles and a critical velocity of 30 m.p.h.

The significance of the above argument may be illustrated by considering a public-transport system. Suppose a given service connects an origin and destination with a desire line of $T$ trips per hour. If the capacity of the vehicle unit is $C$ then the number of vehicle services per hour to meet the demand is $T/C$. Now if the length of the route is $d$ and the speed of the vehicles $v$ the time taken to complete the trip is $d/v$.

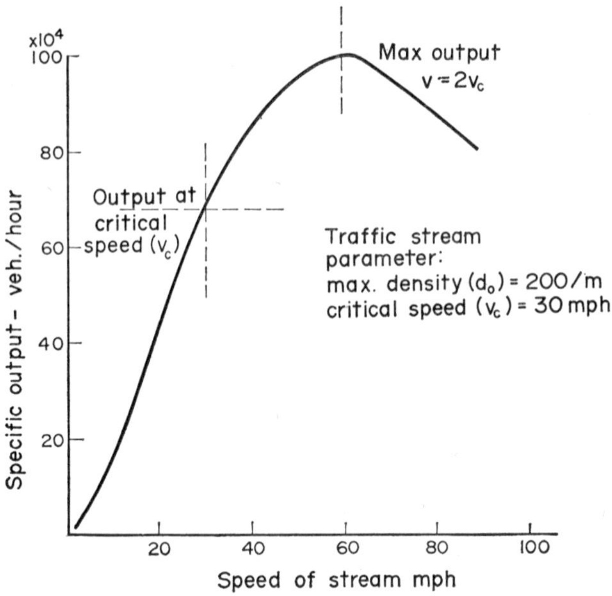


FIG. 5.14. Specific output—vehicle miles/unit time—for transport systems.

The turn-around rate is thus $v/2d$ and the number of vehicles required is

$$N = \frac{2Td}{Cv}.$$

It can be seen, therefore, that the higher the speed the less the number of vehicle units and consequently the lower the fixed costs. However, the operating cost will rise with increase in speed and as a result there will be a characteristic balance point for the system which will yield optimum investment condition. It is not at first obvious that the same situation applies with private cars as they are parked and left idle after the completion of a trip. However, if one considers that there is a fixed number of hours in the year during which a road user desires to devote to driving around, then the criterion is apposite.

The above argument leads to the general proposition that a good index of operational efficiency of a transport link or system is its operating speed. If the speed is higher then it simply does more work.

CHAPTER 6

# Growth and Distribution of Land-use Potential

## INTRODUCTION

Although the overall growth of the potential of a city or regional area is not unrelated to its transport status in both the national or international sense—for example migration to Canada has been over the years greater than to Australia—it is generally necessary in order to confine our analytical procedures within manageable bounds to make estimates of city and regional population and economic growth on a real time-trend basis—modified perhaps by certain expected political and economic inputs. This kind of prediction is primarily for the demographer, economist and politician. In this chapter our concern is with the problem of how an expected total increase in population or jobs will be distributed within city or regional boundaries. Attempts to reduce this kind of problem to analytical form are quite new. Clark(58) may be cited as one of the earliest pioneers of the scientific approach to urban population distribution. There has been also interesting recent work in the application of activity analysis to location theory more particularly in the area of regional planning and industrial location. At this point in time, however, conventional urban land-use planning is more vividly coloured by the artist's brush than the analyst's methodology.

Whilst boundary conditions are necessary to obtain a real solution to any equation, it is vital to distinguish clearly between the constraints and variables. If a theory of land use and transport has any validity, then transport clearly is a major determinant of land-use activity. Although it would, of course, be unfair to say that the town planner is unaware of this important fact, his accounting for it has been essentially subjective. Even the more recent transport plans for cities large and small have been predicted on a single predetermined land-use plan. A notable exception is the current urban renewal and development project for Singapore.

In their turn the transport specialists have been equally myopic in seeing the gravity model as useful only for devising a pattern of desire lines after the trip ends (i.e. the land-use activities) have been fixed rather than as the model for "fixing" the trip ends in the first place. We shall see later that there is an extremely important dual relationship between land-use and desire-line distribution.

The student, however, must first of all trace the history (albeit a short one) of land-use distribution theory.

## EARLY WORK

Because the scientific study of a problem of this kind is not amenable to the simple hypothesis and controlled experiment approach, regression models have provided the main basis for the elucidation of some of the factors likely to influence the different growth patterns. What are the factors that are likely to cause different growth rates in the activity of various land zones? Many come to mind—availability, cost, services, status, views and topography, and in some way transport. Voorhees(59) and his colleagues have both been pioneers and successful exploiters of certain initial findings. Useful current reading on the state of the art with land-use models is given in a paper by Swerdloff and Stowers.(60) The authors test the application of the following models to the growth pattern of the small U.S. city of Greensboro—population approximately 100,000:

A multiple regression model.

The density saturation gradient method.

The accessibility model of Hansen.

Brindle's application of the opportunity models of Schneider and Stouffer.

Although the first is the only true regression model, the others are calibrated by using regression techniques. It should be noted, too, that all models apply to the problem of residential land-use activity.

A review of the early literature reveals that in their enthusiasm to turn the initial models to practical utility, undue emphasis has been placed on calibration to the detriment of evolving theoretical understanding.

The density-saturation gradient model is presented as being essentially empirical, yet it is in fact a version of Clarke's exponential residential density decay function which is in turn an expression of the gravity model concept.

The accessibility model points clearly to the duality between desire-line and land-use density patterns and so also do the two opportunity models. In other words, if given a pattern of trip ends (i.e. residential origins and job destination) the individual worker, in seeking his preferred work zone, is sensitive to the time taken to travel to it, then it is plausible indeed that he may well choose his origin in the first place on account of its proximity to a suitable work place.

Even the regression line model developed for Greensboro demonstrates the overwhelming importance of accessibility. It is instructive to consider the Greensboro equation

$$Y = -2{\cdot}3 + 0{\cdot}061X_1 + 0{\cdot}00066X_2 + 1{\cdot}1X_3 - 0{\cdot}11X_4 - 0{\cdot}0073X_5$$

where $Y$ = logarithm of growth in dwelling units per unit of vacant land, 1948–60,

$X_1$ = zoning protection in 1948 defined by an arbitrary scale 0–9, high values indicated a zoning aimed at preserving a predominantly residential character and low values indicating no zoning protection,

$X_2$ = percentage of the total land in residential use in 1948,

$X_3$ = logarithm of accessibility to employment in 1960,

$X_4$ = dwelling unit density 1948,

$X_5$ = percentage of total land in industrial use 1948.

Before proceeding with further discussion it will be desirable to consider in more detail the accessibility concept.

ACCESSIBILITY

Accessibility of a land use is analogous to the impedance concept of transport capability. It could be thought of in terms of the direct distance, travel time or cost of communicating between an origin and a destination land-use zone. However, if this were done, accessibility would "increase" as the separation between the zones is increased. An inverse function of separation is therefore more appropriate in the descriptive sense and, further, it permits the gravity model impedance function to be exploited.

In general, the accessibility function needs to be extended to take account of multiple origins and destinations and it is plausible and practically convenient to weight the individual accessibility measures by the land-use potential of the set of origin or destination zones as appropriate.

For example, if one wishes to consider the accessibility of a given residential zone, $o$, to a number of work zones $L_d$ $(d = 1, 2, \ldots, n)$ where travel times from $o$ are $t_{od}$, then

$$A_o = \sum \frac{L_d}{Z(t_{od})}.$$

A commonly used form for the accessibility function for work journeys is

$$A_o = \sum_d \frac{E_d}{t_{od}^2} \qquad d = 1 \ldots n, \quad o = 1 \ldots m,$$

where $E_d$ is the number of jobs in work zone, $d$,

$t_{od}$ is the travel time from residential zone, $o$, to work zone, $d$.

EXAMPLE

Develop a table of accessibility indices of the residential zones of ANZAC City in respect of industrial work trips for a future job potential of zones 1, 6 and 8 of 10,000, 11,200 and 1500. It may be assumed that the proposed motorway is in operation.

Perhaps one should observe firstly that the assumed job potential of zones 1, 6 and 8 satisfy the land-use constraints given in Table 1.2,

Chapter 1. And secondly, let it be assumed that the travel impedance function is $Z(t_{od}) = t_{od}^2$. Then in arbitrary units based on measuring employment in 000's and travel time in minutes the accessibility index of zone 2 is

$$A_2 = \sum_d \frac{L_d}{t_{2d}^2} \qquad d = 1, 6 \text{ and } 8$$

$$= \frac{10}{4{\cdot}3^2} + \frac{11{\cdot}2}{14^2} + \frac{1{\cdot}5}{13{\cdot}5^2}$$

$$= 0{\cdot}63.$$

TABLE 6.1. ACCESSIBILITY INDICES FOR INDUSTRIAL AND WHITE-COLLAR WORKERS

| Residential zone | Accessibility indices | |
|---|---|---|
| | Industrial | Commercial and retail |
| 2 | 0·63 | 0·12 |
| 3 | 0·64 | 0·14 |
| 4 | 0·21 | 0·12 |
| 5 | 0·46 | 0·22 |
| 7 | 0·21 | 0·13 |
| 9 | 0·33 | 0·22 |
| 10 | 0·30 | 0·14 |
| 11 | 0·16 | 0·10 |
| 12 | 0·13 | 0·16 |

It is instructive to compare this index with that which would have been obtained without the motorway.

$$A_2 \text{ (without motorway)} = \frac{10}{6{\cdot}3^2} + \frac{11{\cdot}2}{14^2} + \frac{1{\cdot}5}{14{\cdot}5^2}$$

$$= 0{\cdot}32.$$

However, as most of our interest will later on be directed towards ANZAC City of the future we will not pursue the before and after comparisons of this kind. Repetition of the sample calculations yields the second column of Table 6.1 as required. For completeness and use later we will calculate the accessibility of retail and commercial workers. In this case the central area (zone 8) is the only attractor, thus

$$A_o = \frac{L_8}{t_{o8}^2} \qquad o = 2, 3, 4, 5, 7, 9, 10, 11, 12$$

$$\text{and } L_8 = 14{\cdot}3(000\text{'s})$$

The results are tabulated in column 3 of Table 6.1.

## THE ACCESSIBILITY-ORIENTED MODELS

The form of the Hansen Model used in the Greensboro study is

$$G_o = G_T \cdot \frac{A_o^x V_o}{\sum_o A_o^x V_o}$$

where $G_o$ = estimated growth of zone, $o$,
$G_T$ = the total urban or regional growth,
$A_o$ = accessibility index of zone, $o$—as defined above,
$V_o$ = the available land in zone, $o$,
$x$ = experimentally determined exponent.

In the Greensboro calibration $x$ was assessed at about 2·25. Hansen found that in its application to the Washington, D.C., study its optimum value was 2·7. These high values of $x$ only confirm that accessibility is the predominant determinant of residential location.

The two opportunity models are essentially accessibility models, even though the accessibility index is not introduced explicitly. Priority of location is based on the idea that a closer rather than a more distant location will be preferred and the probability of such choice is proportional to the number of opportunities in a given zone to the total number between the zone and the attracting land-use zone—presumably a major work zone or the central area of the city. The opportunity models are most readily applied to the unimodular city, but if accessibility indices with respect of a number of attractor centres are determined for each of the origin zones then the zones may be ranked in order of their accessibility before proceeding with the estimation of locational probabilities.

The two models are known as

(a) The Stouffer model, viz.

$$\delta L_o = G \frac{L_r}{\sum\limits_{o} L_r},$$

$\delta L_o$ = dwelling units forecast to be located in a particular zone $o$,

$L_r$ = opportunities for location in zone $o$,

$\sum L_r$ = total opportunities up to and including zone $o$,

$G$ = a normalizing constant to ensure that total allocation equals the total growth.

(b) The Schneider model may be written

$$\delta L_o = G\{P(\sum_{o=1}^{O} L_r) - P(\sum_{o=1}^{O-1} L_r)\} \quad \text{and} \quad P(L) = 1 - e^{-KL},$$

$K$ = probability of acceptance of a location opportunity. It is a characteristic of the region under study.

The other symbols have been previously defined.

## THE DENSITY-SATURATION GRADIENT MODEL

This method was used as the basis of land-use forecasting in the pioneering of the Chicago Transportation study—the first to exploit the notions of land-use and transport interaction. The density-saturation gradient is, as the name suggests, a plot of the residential saturation as a function of distance or travel time from some convenient reference origin—usually the city centre. It is a function that can be determined experimentally for any city although practical difficulties are encountered because of topographical and zoning peculiarities. It is sometimes necessary to define the function separately for a number of sectors of the surrounding metropolis. Figure 6.1 shows the form of the saturation density function and the associated dwelling unit density function.

One is tempted to the view that whilst such characteristics would readily emerge for a large city, the likelihood of obtaining good definition for small cities or large towns would be somewhat fortuitous.

The results of the Greensboro study did not confirm such doubt. In fact, the essential characteristic of the land-use pattern of community groups—village, town or city, small or large—is the strong nucleation around a central point, be it only the general store and post office at the crossroads.

Colin Clark's extensive researches confirm in a general way this key concept. He has found a negative exponential function fits many of his

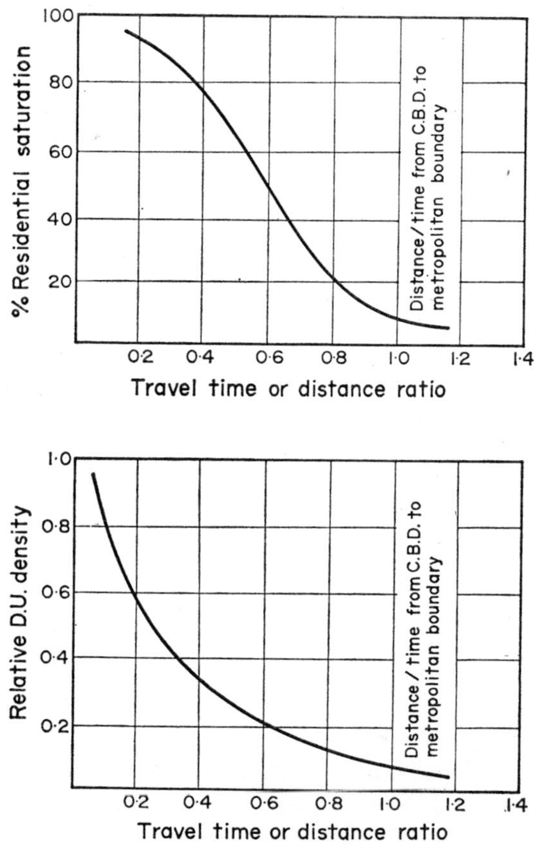


FIG. 6.1. Form of the residential-saturation and dwelling-unit-density characteristics for towns and cities.

data. However, there is scope for much more research as good data becomes available. The gamma function having zero density at the core, rising rapidly to a maximum at 10–20% of the metropolitan radius, then falling relatively quickly to the 50% distance and finally tapering off gradually to the rural hinterland, offers good prospect of exploitation for general theoretical purposes.

## SOME OBSERVATIONS, FUTURE PROSPECTS

So far we have confined our attention to residential land-use growth and in doing this we have emphasized accessibility and explicitly taken account of available land or holding capacity. It should be noted that accessibility may be oriented either to the central area or to all job opportunities. Are there other factors of importance—cost of land, accessibility to recreational facilities, shops, schools, residential amenities, and so on?

What, too, are the prospects for industrial, shopping and school land-use models. If one can accept the dominance of accessibility to jobs in the determination of the pattern of land-use activity a relatively simple philosophy is exposed. The process starts with a more or less arbitrary location of employment activity constrained by land availability, topography and its accessibility to some interregional transport node. Having established centres of employment the residential distribution is shaped by the accessibility function and constrained by holding capacity and price. Once the residential population is located, schools and shopping centres are located at the centroids of residential catchments. Here again accessibility is the prime determinant and population provides the boundary conditions that govern the number and size of such land-use activities.

The location of recreational activities is also primarily influenced by their accessibility to the residential population. For metropolitan areas the "one-off" facilities, i.e. opera houses, major galleries, exclusive entertainment, need central location. Natural recreational land use—lakes, mountains, beaches—cannot of course be "located", but their level of activity depends largely on their transport "distance" from centres of population.

The essential problem to be faced would thus seem to be pointed clearly at the determination of best spatial arrangement of employment and residential zones. As we have seen, the empirical evidence provides good grounds for believing that this arrangement is strongly conditioned by the amount of work-trip travel involved. This in turn suggests that instead of attempting to calibrate physical models, ingenious though they be, programming method may provide a more objective and creative approach to the problem. This is particularly true at the concept planning level where the main objective is to examine the broad implications of the macroscopic pattern of land-use zones and major transport corridors—area of zones, total populations, interzonal distances or travel times.

The study of intrazonal land-use and transport problems involving the optimum distribution of many different land uses, such as high-density and low-density residential, industrial, commercial and institutional, together with their transport connections is also amenable to mathematical programming technology. There is also scope in this more confined context for use of variable oriented systems' models—particularly as their calibration can be more meaningful.

We shall focuse attention therefore on the programming approach, but before doing so we shall illustrate the application of the accessibility model to the ANZAC City.

## A RESIDENTIAL LAND-USE PLAN FOR ANZAC CITY

Most existing accessibility models develop growth coefficient by weighting the holding capacity of the various zones by its accessibility index. A good measure of experience and some current work by Latif in the School of Traffic Engineering shows that land availability is better viewed as a boundary condition or constraint. This is particularly true if the holding capacity is measured in terms of vacant land for if the vacant land diminishes to zero but the accessibility is good, the ultimate effect is to force up the density of land utilization. Conversely in towns and cities large and small there is virtually an infinite amount of vacant land beyond the fringe of its built-up region, but this in itself does not have a strong influence on attracting new residential activity—not at least until the holding capacity of zones, closer to the job centres, is exhausted.

Similarly the price of land which is introduced in some regression models may be introduced as a constraint by stratifying the population by income groups. Data for such stratification is often difficult to obtain and a generally satisfactory alternative is to stratify in terms of job classification at least into white-collar and black-shirt categories. If this is done the residential allocation may be segregated. One should point out that boundary conditions of these kinds are easier to exploit in the programming approach.

In ANZAC City the new residential population to be distributed may be taken as 43,000 and this number is made up essentially of the families of the work force. Real data would provide—down to zonal level—the average family size of the two work categories. We will assume here no difference in average family unit of the two categories of workers. The residential population increment will thus comprise

24,300 — industrial families,

18,700 — commercial and retail families.

If these residents are distributed in the ratio of

$$\frac{A_o}{\sum A_o} \qquad 0 = 2, 3, 4, 5, 7, 9, 10, 11, 12,$$

$$\sum A_o \text{ (indust.)} = 3{\cdot}07, \qquad \sum A_o \text{ (retail and commercial)} = 1{\cdot}35,$$

i.e. the relative accessibility of zone, $o$, then Tables 6.1 and 6.2 result.

It can be seen that the residential populations allocated to zones 5 and 9 exceed their upper limit on holding capacity. We shall therefore allocate to zones 5 and 9 respectively 2400 and 1200 persons and distribute the remaining growth to the other zones, i.e. 2, 3, 4, 7, 10, 11, 12. The problem arises, however, as to how to split the allocations to zones 5 and 9 between industrial and retail and commercial families. This could readily be done by dividing them in the ratio of the increments obtained in the first allocation. On the other hand, it is of practical consequence to point out that in this kind of planning it is often desirable to provide opportunity for planning judgements within the framework of a model which must of necessity ignore many of the real facts of life at the local level. We shall therefore assume that zone 5 is to be redeveloped for higher-income families and zone 9 for the lower-income groups and thus allocate 2400 retail and commercial dependent families to zone 5 and 1200 industrials to zone 9. The second allocation is thus as shown in Table 6.3.

TABLE 6.2. POPULATION INCREMENTS AND NEW ZONAL POPULATION—FIRST ALLOCATION

| Zone | Industrial | Retail and commercial | Old | New |
|---|---|---|---|---|
| 2 | 4930 | 1660 | 3000 | 9590 |
| 3 | 5030 | 1940 | 12,100 | 19,040 |
| 4 | 1670 | 1660 | 3800 | 7130 |
| 5 | 3600 | 3050 | 17,600 | 24,250 |
| 7 | 1670 | 1800 | 1900 | 5370 |
| 9 | 2640 | 3050 | 8800 | 14,490 |
| 10 | 2460 | 1940 | 12,800 | 17,200 |
| 11 | 1270 | 1380 | 2000 | 4650 |
| 12 | 1030 | 2220 | 5000 | 8250 |
| Total | 24,300 | 18,700 | 67,000 | 109,970 |

TABLE 6.3. POPULATION DISTRIBUTION—SECOND ALLOCATION

| Zone | Industrial | Retail and commercial | Old | New |
|---|---|---|---|---|
| 2 | 6400 | 2150 | 3000 | 11,550 |
| 3 | 6450 | 2470 | 12,100 | 21,020 |
| 4 | 2140 | 2150 | 3800 | 8090 |
| 5 | 0 | 2400 | 17,600 | 20,000 |
| 7 | 2130 | 2350 | 1900 | 6380 |
| 9 | 1200 | 0 | 8800 | 10,000 |
| 10 | 3040 | 2500 | 12,800 | 18,340 |
| 11 | 1620 | 1800 | 2000 | 5420 |
| 12 | 1320 | 2880 | 5000 | 9200 |
| Total | 24,300 | 18,700 | 67,000 | 110,000 |

It is interesting to compare the above result with

(a) the Voorhees approach. The holding capacities, $H_o$, can be obtained by differencing the present and maximum columns of Table 1.2, Chapter 1. The allocation can then be made by using the formula

$$G_o = G^T \frac{A_o H_o}{\sum A_o H_o} \qquad o = 2, 3, 4, 5, 7, 9, 10, 11, 12,$$

where $H_o$ is the holding capacity of zone $o$ for each class of worker.

(b) The regression approach based on the LARTS model with coefficients as given by Blunden in *Introduction to Traffic Science*,[(2)] page 15.

TABLE 6.4. COMPARISON OF ESTIMATES OF CHANGE IN THE RESIDENTIAL LAND-USE POTENTIAL FOR ANZAC CITY

| Zone | Accessibility (holding capacity constraint) | Accessibility weighted H.C. | Regression LARTS |
|---|---|---|---|
| 2 | 11,550 | 9260 | 9000 |
| 3 | 21,020 | 19,410 | 17,000 |
| 4 | 8090 | 11,480 | 13,200 |
| 5 | 20,000 | 19,330 | 18,950 |
| 7 | 6380 | 9100 | 12,300 |
| 9 | 10,000 | 9570 | 8100 |
| 10 | 18,340 | 15,150 | 14,150 |
| 11 | 5420 | 6830 | 7700 |
| 12 | 9200 | 9870 | 9600 |

It is left to the reader to work through these methods. The results are compared in Table 6.4. The student would do well to ponder them and raise for debate the relative merits of an approach oriented towards "extrapolating" current behaviour as compared with the opportunity to direct it towards more arbitrary goals. The degree of extrapolation inherent in the variable oriented (or systems) models' depends on how tightly they are calibrated. If calibration is minimal then the model has power as a creative planning tool. If the calibration is severe then they tend towards regression-type predictions.

With this issue before us it is appropriate then to proceed to a discussion on programming or optimizing methods for land-use distribution.

## PROGRAMMING METHODS

In Chapter 4 a simple example was given of the application of the classical transportation problem to determination of desire-line patterns by minimizing the total amount of travel in the $O$ and $D$ system. The

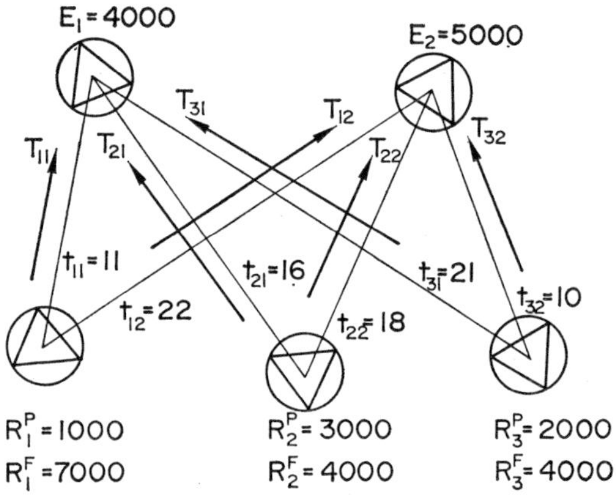


FIG. 6.2. Land-use and transport data for optimization of residential locations of work force.

dual of this problem defines a set of variables, $O_1$, $O_2$, $O_3$, $D_1$, $D_2$, which are identifiable with the level of land-use activity of the origin and destination zones. They represent imputed values or opportunity costs or marginal variations in the zonal activities and these values or costs are measured in terms of travel time, distance or cost. The marginal trends thus exhibited prompt one to examine the consequences of relaxing the zonal constraints by non-marginal amounts and this in turn suggests the setting of a lower bound equal to the present level of activity and an upper value corresponding to the maximum desirable level. In the context of land-use development, the lower and upper bounds are clearly existing population and holding capacity.

In general we have three options, viz.

(a) fix jobs and relax residential constraints,
(b) fix residential and relax job constraints,
(c) relax both job and residential constraints.

We have explained already that the distribution of employment centres is constrained by a number of exogenous factors and policies and it will serve our purpose here to extend the example of Chapter 4 as a residential land-use model. Figure 6.2 illustrates the general situation.

The travel-time matrix may be taken as that used previously. The objective function may be written as before

$$Z = 11T_{11}+16T_{21}+21T_{31}+22T_{12}+18T_{22}+10T_{32}$$

and this we desire to minimize subject to the following land-use constraints:

$$\begin{aligned}
&T_{11}+T_{21}+T_{31} &&= 4000\\
&T_{12}+T_{22}+T_{32} &&= 5000\\
&T_{11}+T_{12} &&\geqslant 1000 \leqslant 7000\\
&T_{21}+T_{22} &&\geqslant 3000 \leqslant 4000\\
&T_{31}+T_{32} &&\geqslant 2000 \leqslant 4000
\end{aligned}$$

with all $T$'s $> 0$.

It should be noted that the slack variables taken together with the artificial variables for the first two equations do not constitute an initial basis. As a consequence additional artificial variables are needed for the "greater than" constraints. The simplex algorithm yields the following result:

$$\begin{aligned}
T_{11} &= 2000 \qquad & T_{21} &= 2000 \qquad & T_{31} &= 0\\
T_{12} &= 0 \qquad & T_{22} &= 1000 \qquad & T_{32} &= 4000
\end{aligned}$$

This set of trip interchanges determines uniquely the distribution of the residential trip ends, viz.

$$R_1 = 2000 \qquad R_2 = 3000 \qquad R_3 = 4000$$

The employment ends have, of course, been specified in the data.

The value of the object function is $Z = 112{,}000$ person minutes. If one compares this result with the illustrative example of the transportation problem in Chapter 4 it will be seen that the constraints on the residential land-use distribution have resulted in less travel (i.e. 3000 person minutes). Further, if the lower bound for $R_2$ were reduced to zero and the upper bound of $R_3$ increased to 5000 then the solution degenerates* to $T_{11} = 4000$ and $T_{32} = 5000$ with other $T$'s $= 0$. The total travel is then 94,000 person minutes. It should be noted, however, that even though the problem is set up to determine a desire-line distribution the land-use pattern is implicit in the solution.

The dual formulation leads to the coefficients of the marginal travel matrix. It is left to the student to verify its formulation, viz.
Maximize

$$W\dagger = 4 \,.\, D_1 + 5 \,.\, D_2 + O_1^P - 7 \,.\, O_1^F + 3 \,.\, O_2^P - 4 \,.\, O_2^F + 2 \,.\, O_3^P - 4 \,.\, O_3^F$$

subject to

$$D_1 + O_1^P - O_1^F = 11 \qquad D_2 + O_1^P - O_1^F = 22$$
$$D_1 + O_2^P - O_2^F = 16 \qquad D_2 + O_2^P - O_2^F = 18$$
$$D_1 + O_3^P - O_3^F = 21 \qquad D_2 + O_3^P - O_3^F = 10$$

where the superscripts $P$ and $F$ refer to the present and future populations of the residential zones 1, 2, 3. In practice it is preferable to redefine the population constraints within the present and future bounds.

The above set of equations can be solved in the form given with the following result:

$$D_{11} = 11 \qquad D_2 = 13 \qquad O_2^P = 5 \qquad O_3^F = 3$$
$$O_1^P = O_1^F = O_2^P = O_3^F = 0$$

* Degeneracy in the L.P. context is a matter of considerable theoretical and computational importance. A basic solution is degenerate if one or more of the basic variables are zero. In the transportation problem there are $(n+m-1)$ basic variables. In this example $n+m-1 = 4$. Degeneracy does not invalidate the solution of the primal but introduces complications into the dual solution. It is not proposed to pursue this matter in this introductory account of L.P. methodology. The student is referred to the standard texts.[38–40]

† Coefficients of objective function measured in 000's.

In Chapter 4 the marginal travel matrix was defined in terms of the dual variables. When we turn now to the practical implications of this matrix it is appropriate to subdivide it into a pair of matrices—for origins and destinations. This is done because no real meaning can be attached to interchanging a residential origin for a factory destination—unless of course the worker sets up his bed in his office! We have thus two matrices

TABLE 6.5. THE MARGINAL TRAVEL MATRIX

(a) *Marginal travel matrix for origins*

| From | To | | | |
|---|---|---|---|---|
| | $R_1$ | $R_2$ | $R_3$ | $Z_o$ |
| $R_1$ | | 5 | −3 | 0·66 |
| $R_2$ | −5 | | −8 | −4·33 |
| $R_3$ | +3 | +8 | | 3·66 |

(b) *Marginal travel matrix for jobs*

| From | To | | |
|---|---|---|---|
| | $E_1$ | $E_2$ | $Z_d$ |
| $E_1$ | 0 | −2 | −1·00 |
| $E_2$ | +2 | 0 | 1·00 |

$Z_o$ or $Z_d$, the zone index for origins or destinations is given in the last column and is simply defined as

$$Z_o \text{ (or } Z_d) = \frac{1}{o} \sum_{x=1}^{o} M_{ox} \qquad x = 1, 2 \ldots o.$$

It is useful in large tables as it is tedious and difficult to trace the effect of every interzonal marginal change. For example, in Table 6.5(a) it can be seen at a glance that moving residential locations out of zone $R_2$, irrespective of where, gives a rather large reduction in travel whereas moving a residence from $R_1$ or $R_3$ has the effect of increasing travel. It can also be seen that $R_2$ is more sensitive to relocation activity.

One should observe that instead of holding the job distribution constant and seeking an optimum allocation of residential locations the converse problem could be solved. Further, it is possible to relax the constraints on both job and residential locations and seek a truly optimal distribution of land uses.

### EXAMPLE

Consider a land-use/transport system geometrically similar to that discussed above. Now suppose that the expected future population of each residential zone is to be 3000 workers. The present number of jobs in the employment centres $E_1$ and $E_2$ is 2000 and 3000 respectively and there is growth potential for a total of 7000 and 5000. How should an additional 4000 jobs be allocated to minimize the amount of travel in the system? Show that the total amount of travel for this optimum condition would be 111,000 travel units.

### APPLICATION OF THE L.P. FORMULATION TO ANZAC CITY

In applying the linear programming method to ANZAC City a number of cases arise, viz.

1. Job Locations Fixed, Jobs and Workers unstratified.
2. Worker Residences Fixed, Jobs and Workers unstratified.
3. Job Locations Fixed, Jobs and Workers stratified.
4. Worker Residences Fixed, Jobs and Workers stratified.
5. Job and Worker Residence not Fixed, Jobs and Workers stratified.

We shall consider case 3 in some detail. Before doing so it is appropriate to explain briefly the meaning, in this context, of *stratification*. Stratification is the subdivision of a group of trips into more homogeneous sub-groups. The common breakdown of trips by purpose is a form of stratification of the total daily trips. However, it is clear that a great deal of information is hidden if work trips are defined only in terms of their length and direction. Important sub-classifications of work trips include

industrial,
commercial and administrative,
retail,
male,
female,
executive.

It is of some interest to note that the need for finer detail in setting up a problem is closely correlated with the sensitivity or resolving power of the tools of analysis—the more sensitive the greater the need for detail. Linear programming is a most discriminating tool and when applied to problems involving travel patterns a fine stratification is necessary. The gravity model, on the other hand, in its role as a trip distributor, is relatively insensitive to absolute values of the variables and as a result can deal effectively with heterogeneous data.

In proceeding then with the land-use plan for ANZAC City we will stratify the work trips and their origins and destinations into industrial and white-collar categories.

The stratification of the jobs is clear from Table 1.2. In the residential zones the existing distribution of industrial and white-collar workers would normally be available from data. The availability of accommodation for additional blue- and white-collar workers is arbitrarily set.

The land-use distribution model may thus be written. Minimize

$$Z = \sum_{o=1}^{O} \sum_{d=1}^{D} T_{od}^{I} t_{od} + \sum_{o=1}^{O} \sum_{d=1}^{D} T_{od}^{C} t_{od}$$

with the following constraints:

1. $$\sum_{o} T_{od}^{I} = E_{d}^{I} \qquad d = 1, 6, 8,$$

2. $$\sum_{o} T_{od}^{C} = E_{d}^{C},$$

3. $$\sum_{d} T_{od}^{I} \geqslant R_{o}^{I_P} \qquad o = 2, 3, 4, 5, 7, 9, 10, 11, 12,$$

4. $$\sum_d T_{od}^I \leqslant R_o^{I_f},$$

5. $$\sum_d T_{od}^I \geqslant R_o^{C_p},$$

6. $$\sum_d T_{od}^C \leqslant R_o^{C_f},$$

7. $$\sum_l \sum_d T_{ld}^I = R_L^I \quad l = 2, 3, 4, 7,$$

8. $$\sum_l \sum_d T_{ld}^C = R_L^C, \quad \text{and all } T\text{'s} \geqslant o$$

where $E_d^I$ and $E_d^C$ are the predicted number of industrial and residential jobs in the three work zones 1, 6 and 8;

$R_o^{I_p}$, $R_o^{I_f}$, $R_o^{C_p}$ and $R_o^{C_f}$ are the existing and maximum populations of the zones for industrial and commercial workers respectively, i.e. residential zones 2, 3, 4, 5, 7, 9, 10, 11 and 12;

$R_L^I$ and $R_L^C$ are the total numbers of industrial and commercial workers who must of necessity live in low cost residential areas, i.e. zones 2, 3, 4 and 7.

We have already made arbitrary estimates of the future job distributions, viz.

$E_1^I$, $E_6^I$ and $E_8^I$ are 9,200, 11,600 and 1,900,

$I_1^C$, $E_6^C$ and $E_8^C$ are 0 0 14,300.

The residential constraints are given in Table 1.2. However, total population figures are shown. It is necessary to convert these to worker residents and stratify them by job class. In practice there would be much real data available for the development of these constraints. Here we arbitrarily scale the total population in the ratio of industrial and commercial worker categories. Table 6.6 gives the result.

Table 1.2 also shows that residential zones 2, 3, 4 and 7 are low-cost areas. We may assume arbitrarily that 10,600 industrial and 5000 commercial workers will require low-cost living areas. We are now in a position to develop the above system of equations. Consider, for example,

Equation 1 which yields for industrial trips destined for zone 1—

$$T^I_{21}+T^I_{31}+T^I_{41}+T^I_{51}+T^I_{71}+T^I_{91}+T^I_{10,1}+T^I_{11,1}+T^I_{12,1} = 9200.$$

Equations 3 and 4 give for outbound work trips from zone 2—

$$T^I_{21}+T^I_{26}+T^I_{28} \geqslant 600,$$

$$T^I_{21}+T^I_{26}+T^I_{28} \leqslant 2700.$$

TABLE 6.6. EXISTING AND MAXIMUM DISTRIBUTION OF WORKERS' RESIDENCES

| Zone | Resident workers | | | |
|---|---|---|---|---|
| | Industrial | | Commercial | |
| | Existing | Max. | Existing | Max. |
| 2 | 600 | 2700 | 400 | 1300 |
| 3 | 2700 | 4100 | 1400 | 2400 |
| 4 | 800 | 5500 | 500 | 4000 |
| 5 | 3700 | 4000 | 2200 | 2500 |
| 7 | 500 | 5500 | 200 | 4000 |
| 9 | 1900 | 2000 | 1000 | 1500 |
| 10 | 2700 | 4400 | 1500 | 2000 |
| 11 | 500 | 4500 | 200 | 3700 |
| 12 | 1100 | 4300 | 600 | 3600 |

Equation 7 indicates that 10,600 members of the industrial work force must be accommodated in low-cost zones, viz.

$$\sum_d T^I_{2d}+\sum_d T^I_{3d}+\sum_d T^I_{4d}+\sum_d T^I_{7d} = 10,600.$$

$$T^I_{21}+T^I_{26}+T^I_{28}+T^I_{31}+T^I_{36}+T^I_{38}$$

$$+T^I_{41}+T^I_{46}+T^I_{48}+T^I_{71}+T^I_{76}+T^I_{78} = 10,600.$$

Colston[11] has taken out a solution of this and the other four cases. The results of case 3 are summarized in Table 6.7.

From this table it is easy to compile the new zonal populations by summing the industrial and commercial jobs and expanding the sum by a factor of 3. The final results are given in Table 6.8. Colston's solutions to this and the other four cases are illustrated in Figs. 6.3, 6.4, 6.5, 6.6 and 6.7.

It is instructive to compare Table 6.8 with Table 6.4. There is, of course, a good general similarity between the patterns obtained by all methods, particularly when set against the uncertainties that are bound

TABLE 6.7. ZONE TRIP ENDS AND MARGINAL TRAVEL INDICES

| Zone | Zone trip ends | | | |
|---|---|---|---|---|
| | Industrial | *Z* | Commercial | *Z* |
| Destination | | | | |
| 1 | 9200 | −1·67 | 0 | −1·00 |
| 6 | 11,600 | 3·33 | 0 | −2·17 |
| 8 | 1900 | −1·67 | 14,300 | 3·17 |
| Origin | | | | |
| 2 | 2700 | −0·07 | 1300 | 0·08 |
| 3 | 4100 | 0·54 | 2400 | 0·66 |
| 4 | 800 | −3·47 | 500 | −0·34 |
| 5 | 4000 | 3·95 | 2500 | 1·16 |
| 7 | 1600 | −2·20 | 2200 | 0·06 |
| 9 | 2000 | 4·44 | 1500 | 1·16 |
| 10 | 4400 | −3·04 | 2000 | 0·50 |
| 11 | 500 | −3·16 | 200 | −2·94 |
| 12 | 2600 | −3·06 | 1700 | 0·34 |

TABLE 6.8. RESIDENTIAL ZONE POPULATIONS

| Zone | Total workers | Population |
|---|---|---|
| 2 | 4000 | 12,000 |
| 3 | 6500 | 19,500 |
| 4 | 1300 | 3900 |
| 5 | 6500 | 19,500 |
| 7 | 3800 | 11,400 |
| 9 | 3500 | 10,500 |
| 10 | 6400 | 19,200 |
| 11 | 700 | 2100 |
| 12 | 4300 | 12,900 |

to influence long-term planning exercises of this kind. However, the programming approach highlights accessibility considerations as the slow growth of zones 4 and 11 shows. In doing so it agrees best with the relative accessibility model (col. 2, Table 6.4). One is tempted to observe

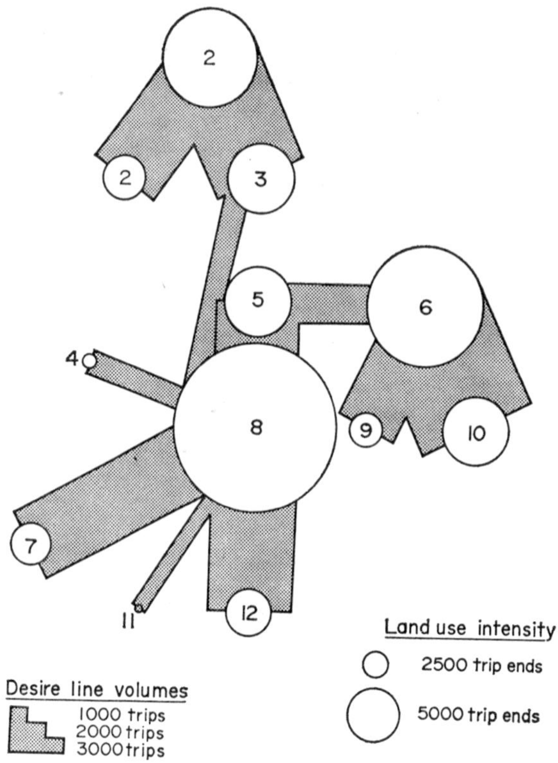


FIG. 6.3. Linear programming solution model 1.

that we have here a method of combatting the problem of urban sprawl! On the other hand, if for various reasons there is need to encourage development in certain more remote "empty" areas then this can be taken into account in the programming approach by introducing an appropriate constraint.

## OTHER FORMS OF THE OBJECTIVE FUNCTION

In the foregoing discussion we have focused attention on the minimization of work travel and used this as the objective function in the programming methods. There is a great deal of empirical evidence to

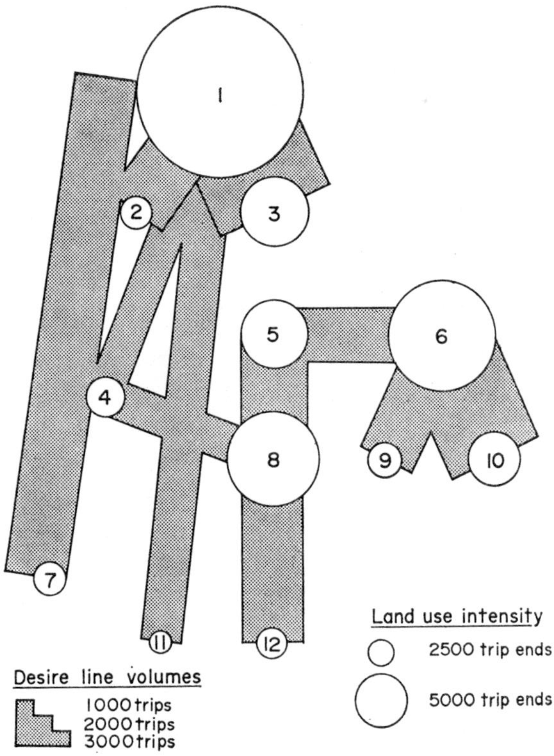


FIG. 6.4. Linear programming solution model 2.

support the contention that, subject to certain boundary conditions and constraints, proximity to work is the dominant factor influencing residential location. Not only is the desire to minimize work travel a plausible criterion, so far as the individual is concerned, it is also a most desirable objective for government and civic authorities, because the costs of road building and subsidization of other transport facilities is not only a major item in their budgets but because also the scale of the works or operations is largely determined by the peak-hour traffic demand, i.e. the work trips.

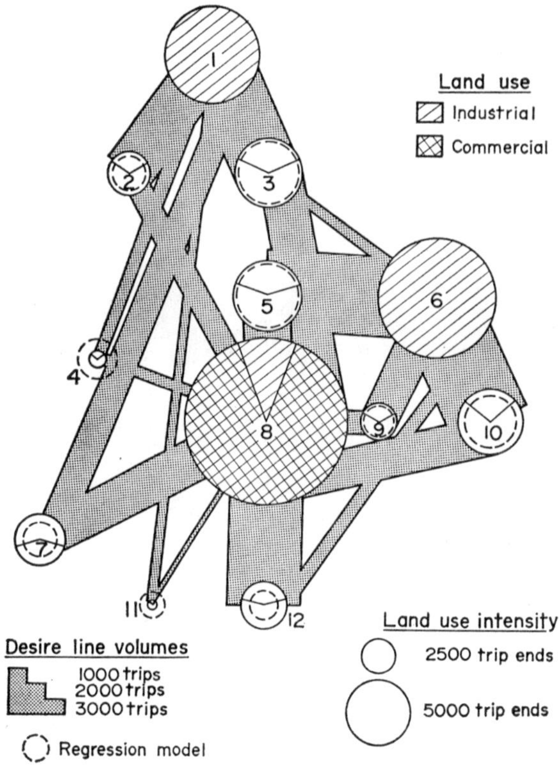


FIG. 6.5. Linear programming solution model 3.

There are, however, other criteria for work travel that need to be mentioned and investigated by the new generation of students.

(a) Minimization of passenger miles of travel.
(b) Minimization of travel costs.
(c) Minimization of a linear combination of travel and locational costs.

The first of these is useful in concept planning of a land-use activity pattern even though the performance of the transport system under traffic loading is ignored.

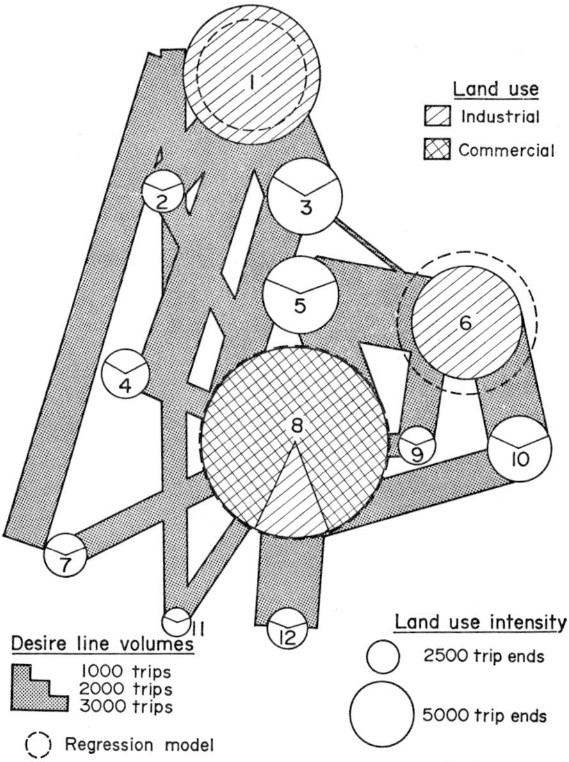


FIG. 6.6. Linear programming solution model 4.

Although there are some advantages in rating traffic activity in economic terms there are complications as well. Taken overall not much is gained in the general land-use planning context because in urban conditions costs are proportional to total travel time to a good order of approximation. If time is used as the transport impedance measure one avoids the difficulty of specifying a money value for the private time component of the transport cost. In a recent paper, Goodwin (unpublished), of the Department of Traffic Studies, University College, London, shows that the time value of money has better correlation

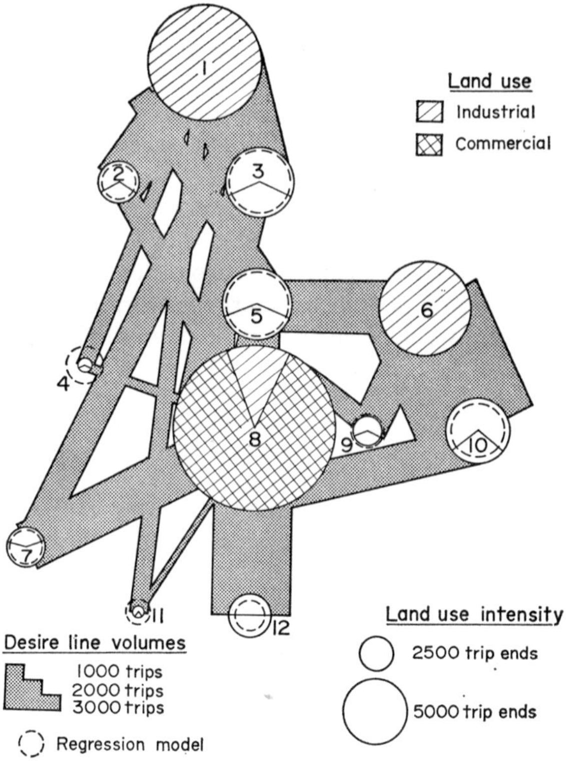


FIG. 6.7. Linear programming solution model 5.

with wage rates than does its reciprocal. This is most interesting because it permits one to "adjust" the travel time by adding an increment obtained by converting the actual operating or out-of-pocket cost component into a time measure.

The last case raises some interesting possibilities. One could introduce a coefficient of the form $(C_{od}+C_o)$ where

$C_{od}$ = annual travel cost/unit on link *od*,

$C_o$ = annual cost index of a residential unit of the origin land-use zone,

for use in the objective function. Rough estimates suggest that $C_{od}$ and $C_o$ are of comparable but complementary magnitude for given work-force strata. The objective function will thus exercise little discrimination in the determination of the land-use activity pattern. The student might ponder over this observation and in the light of the strong empirical evidence that supports the inverse travel-time models ask himself whether or not the sociological implications of "proximity" are not in fact of greater significance than pure economic considerations.

If we turn to non-work trips other criteria merit consideration. For recreational trips it would seem appropriate to maximize some function that measures travel desire, viz.

$$Z = \sum_o \sum_d \frac{L_d}{Z(t_{od})} \cdot T_{od}$$

where the $L_d/Z(t_{od})$ may be regarded as the travel-desire coefficient of recreational centre, *d*, with respect to the various residential locations. The maximization of the travel desire would be subject to transport system capacity constraints and trip-end or travel limitations.

In the case of shopping trips or shopping-centre location the optimization tends towards maximization rather than minimization. It is not too difficult to establish by heuristic argument that maximization of total trips is a plausible proposition—people "like to go shopping". In addition to the obvious constraints it would be appropriate here to consider trip-length or travel-time restraints.

CHAPTER 7

# Traffic Generation

## INTRODUCTION

We have already explained in Chapter 1 that traffic generation is a unique measure of land-use activity and the traffic generated by a land-use zone or area may be regarded as a flux in much the same sense as, say, the magnetic flux of a piece of iron. We have explained also that as yet no general theories of this complex phenomenon have emerged.

However pressed by the exigencies of current "traffic problems", the traffic engineer has in recent years assembled a great amount of data on the traffic generated by various types of land use. Whilst little use has been made of it for theoretical purposes, it has been extensively used for a variety of traffic-planning purposes. In this sense the traffic engineer has been little different from his electrical engineering counterpart. To refer again to our analogy with magnetism, one might point out that the lack of theoretical formulae to predict the magnetic effects of a bar of iron or special magnetic alloy has not prevented their determination by direct measurement. The magnetization curve thus obtained provides the necessary data for the design of all kinds of magnetic and electromagnetic devices. Perhaps it is well to point out that a great number of the important physical constants used by the engineer can only be determined by experimental measurement—ultimate tensile strength of materials, yield points, conductivity, coefficients of expansion, and so on.

However, in spite of the apparent adequacy of experimental constants, the physicist and other scientists are always busy seeking new relationships that will explain observed phenomena more completely and describe their characteristics in more fundamental terms. This same process is under way in the area of land-use phenomena and in spite of the novelty of the research some rationale is being established.

We have explained also that the traffic generation flux is proportional to the traffic potential of the land use and we have defined the flux density across the perimeter of its boundary contour. It is appropriate now to give consideration to the time and space dependency of this traffic flux.

### *Intensity of Traffic Generation*

The traffic generation of land usage varies of course from place to place and from moment to moment. It is necessary therefore to define its intensity as a function of space and time coordinates, i.e. $x$, $y$ and $t$. If the number of trips emanating from an elementary land-use area $\delta x \, . \, \delta y$ in time $\delta t$ is $\delta T_o$ and the intensity of traffic generation is $g(x, y, t)$, then

$$\begin{aligned} \delta T_o &= g(x, y, t)\,\delta x\,\delta y\,\delta t \\ &= g(A, t)\,\delta A\,\delta t. \end{aligned} \tag{7.1}$$

The basic measures of traffic generation are persons or tons. Although it is appropriate in many circumstances to measure the traffic generation in terms of cars, trains, planes, ships and so on it is essential if we are to pay more than lip service to the cause of comprehensive transport planning to think first in terms of persons and tons and later on convert to the other units by introducing the vehicle capacity or occupancy. The measures of traffic generation intensity are thus

| | |
|---|---|
| persons | per unit area per unit time, |
| tons | per unit area per unit time. |

In many situations it is common to specify the land use descriptively—a given zone, a supermarket, hospital, service station—and the area unit may in consequence be dropped. Nevertheless, it should be emphasized that the space intensity of traffic generation varies by an order of magnitude if one compares, say, a parking lot and a ten-storey parking structure of similar capacity. Of particular importance is the dependence of traffic generation intensity on time—particularly as it gives rise to the problems of peak-and rush-hour traffic.

It should be noted here that even though the exploitation of the Average (or Annual Average) Daily Traffic measure has many useful applications in rural road planning, its adaptation to the urban scene has done much to inhibit the understanding of many of the special problems of urban transport activity—but more on this issue later on.

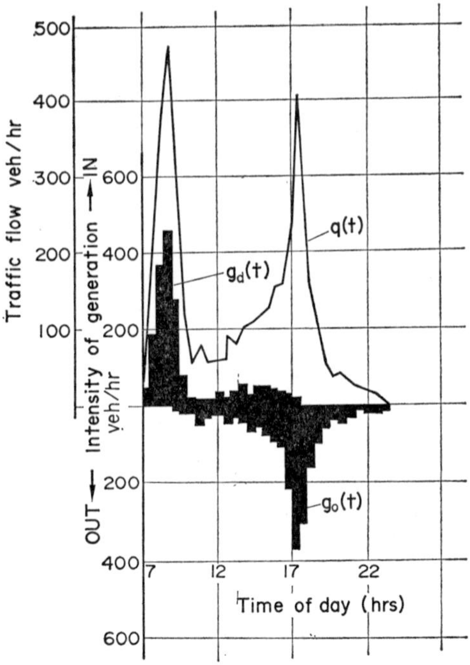


FIG. 7.1(a). Generation intensities and traffic flow for large commuter parking station on fringe of C.B.D., Sydney (Goulburn Street)—capacity 750.

Figure 7.1 illustrates the number of salient points concerning traffic-generation intensity and certain associated measures—trip ends and flow rates. The data presented is taken from a careful analysis of parking tickets issued at a large parking station on the fringe of the central business area of Sydney. It is essentially a commuter and businessman's carpark but because of its premium location a small queue of hopefuls continuously exists during the off-peak period waiting to take advantage of vacancies created by the small turnaround of businessmen's vehicles.

Figure 7.1(a) plots the entries and exits averaged over a half-hour interval as a function of time of day—7 a.m. to 10 p.m. If the parking

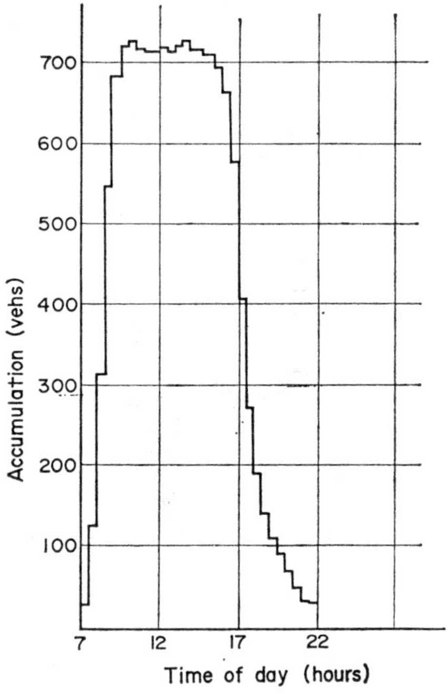


FIG. 7.1(b). Accumulation characteristic—Goulburn Street parking station, Sydney.

station is regarded as a destination or a sink then the intensity of traffic generation is given by the positive values. Considered as origin of trips or a source, then the lower curve is the intensity function. It will be noted that the intensity varies from a few vehicles per hour to a maximum 465/hour inbound or 380/hour outbound.

If the number of traffic units arriving at or departing from a given lane-use zone in a small interval of time $\delta t$ is $N_r$, then the number arriving in a finite time made up of $n$ small units is

$$T = \sum_{r=1}^{n} N_s \quad \text{and} \quad n\delta t = T \tag{7.2a}$$

or in terms of the intensity function $g$

$$T = \int_{t_1}^{t_2} g_d(t)\,dt \quad \text{or} \quad \int_{t_1}^{t_2} g_o(t)\,dt, \tag{7.2b}$$

where $g_d$ and $g_o$ are the intensity functions when the land use is acting as a destination or an origin, respectively, and $T$ the trip ends. The integral of the difference between these intensity functions gives the accumulation, i.e.

$$A = \int_{t_1}^{t_2} \{g_d(t) - g_o(t)\}\,dt. \tag{7.3}$$

This characteristic is shown in Fig. 7.1(b).

The flow rate (to and from) the land use is the sum of the generation intensities,

$$q(t) = g_d(t) + g_o(t). \tag{7.4}$$

Figure 7.1(a) leaves no doubt that it is a highly time dependent phenomenon. More generally we may write the total traffic generation, i.e. the total trip ends as

$$T = \int_x \int_y \int_z g(x, y, t)\,dx\,dy\,dt. \tag{7.5}$$

It should be stressed that trips are a vector-like quantity and should be related to an origin *or* a destination. In this sense the ADT concept of traffic desire lines is particularly hazardous.

Although it is possible to integrate the intensity function to give a trip-end pattern for a continuum of land-use activities the basic data required to specify its functional form is in general unavailable. It is instructive to use certain hypothetical functions for analytical exercises. One profitable use of the underlying concepts' relates to the determination of the "centroid" of traffic activity of an extended land-use zone. The following examples will illustrate these points.

EXAMPLES

1. Consider a circular residential zone of 1 mile diameter. The density of housing falls off exponentially from the centre outwards and has a value of 7 per gross acre in the central region. If the number of workers whose jobs are outside the zone is 1·5 per dwelling unit, determine the origin work trip ends that should be represented at the zone centre. If there are ten effective outbound road lanes at the boundary, what is the density of the traffic generation flux in vehicles per lane, assuming the vehicle occupancy is 1·5?

The housing density may be represented by

$$d = De^{-r/R}.$$

Here $D = 7/\text{acre} = 4{\cdot}48 \times 10^3/\text{sq. mile},$

$$R = \tfrac{1}{2} \text{ mile}.$$

Therefore $d = 4{\cdot}48 \times 10^3 e^{-2r}.$

The number of houses in an annular ring distance $r$ from the centre is

$$N = 2 \times 4{\cdot}48 \times 10^3 r e^{-2r} dr,$$
$$28{\cdot}2 \times 10^3 r e^{-2r} dr.$$

The traffic generation flux emanating from the ring is thus

$$T = K.N = 1{\cdot}50N$$
$$= 42{\cdot}3 \times 10^3 r e^{-2r} dr.$$

The total traffic generation during the work trip period is

$$T = 42{\cdot}3 \times 10^3 \int_0^{r=\frac{1}{2}} r e^{-2r} dr$$

$$= 42{\cdot}3 \times 10^3 \left[\frac{e^{-2r}}{2^2}(-2r-1)\right]_0^{\frac{1}{2}}$$

$$= 42{\cdot}3 \times 10^3 \times \tfrac{1}{2}\left(\tfrac{1}{2} - \frac{1}{e}\right)$$

$$= 29{,}600 \text{ persons/peak period.}$$

The traffic generation in car units is

$$T_c = 19{,}700 \text{ cars}$$

and the traffic flux density is

1970 per lane.

2. What would be the total traffic generation in the peak period if the zone were semicircular but in other respects conditions are the same as in Example 1? Where would you locate the zone centroid?

The first part of the question is answered simply—it is just half the value obtained above, i.e. 14,800 person trips.

The second part of the question is included to show how quickly analytical tools fail in many practical problems. If the distribution of land-use potential were uniform the centroid would be a distance $4R/3\pi$ from the centre along the radius of symmetry, i.e. in this case 0·21 miles. In this particular example it would clearly be closer but an analytical solution is intractable. However, it can be determined, with difficulty, numerically and the location is approximately 0·15 miles from the centre of the circle.

In practical problems the best way to go about the problem of determining centroids is clearly by specifying the Cartesian coordinates of a number of sub zones of uniform land-use potential and determining the centroid in the usual manner.

### *Vehicle Capacity and Load Factor*

A discussion of vehicle occupancy or capacity has already been given in Chapter 5. It is important here to emphasize that whilst the basic units of traffic generation are persons and tons the traffic flow it creates, must be obtained by taking due account of the capacity and load factor of the transport vehicle.

This is, of course, a number of situations where the car is for all practical purposes the basic unit for measuring traffic generation. Notably amongst these are the various "drive-in" land uses—supermarkets, theatres, parking stations, service stations, motels, food and drink retailing facilities and many others. More will be said concerning these specialized land uses later on.

### *Methods of Determining Traffic Generation*

There are four main approaches to the problem of determining the traffic generation of a land-use zone. They are:

(a) Direct Measurement.
(b) Regression Line Models.
(c) Functional Models.
(d) Parameters of the Land-use/Transport Interactance Models.

#### DIRECT MEASUREMENT

A vast amount of data has been accumulated on the gross traffic generation of land-use zones. It is relatively easy to obtain such data from the cordon counts that have been a feature of very many transportation studies. As a result the data has been obtained for relatively large zones or aggregations of zones. In fact the internal/external survey yields data on the traffic generation characteristic of a whole metropolis. More often the studies of this kind have been focused on the central business district. A simple cordon count on the boundary of a large zone such as the CBD does not, of course, reveal the through-traffic component. On the other hand, from the point of view of traffic generation of a given land-use zone there is no reason to exclude through traffic even though its presence within the zone is short lived. Accumulation which is the difference between in and out flows is not markedly affected by the inclusion of through traffic. In the sense that the flow across the cordon is a flux then it is immaterial whether or not some of it is through traffic.

On the other hand, the matter of through traffic is basic to the whole concept of transport and land-use planning, for the trunk transport network should be designed both to service just this kind of traffic and at the same time ensure that the local land-use functions are not compromised by it roaming freely over the internal road system.

The direct measurement method is more efficacious when applied to zones which have a high traffic flux, i.e. the CBD, industrial estates, individual land uses such as supermarkets, regional centres, parking stations and so on. It is important, however, to measure the intensity of the traffic generation at least to the extent of defining the peak generation characteristics. One of the most notable direct measurement studies of the city centres in Britain was carried out in 1968 by the Road Research Laboratory.(62) From the data obtained many simple and useful functional relationships were obtained, viz.

(a) Area of central area (sq. miles) $= 8{\cdot}8 \times 10^{-5}(P)^{2/3}$ (7.6a)
where $P$ is the metropolitan population.

(b) Peak flow $(q) = 6 \,.\, P^{0{\cdot}6}$ p.c.u./hour. (7.6b)

(c) Bus flow $= 2 \times 10^{-3} \,.\, P$ buses/hour. (7.6c)

(d) Peak flow per unit length of cordon

$$q = 14{,}000 \,.\, A. \qquad (7.6d)$$

(e) Accumulation of persons and vehicles

$$\text{persons} — 0{\cdot}07 \,.\, P, \qquad (7.6e)$$

$$\text{vehicles} — 0{\cdot}76 \,.\, P^{0{\cdot}7}. \qquad (7.6f)$$

(f) Intermodal assignment (bus vs. cars)

$$\left.\begin{array}{l} 75\% \text{ by bus for cities } > 10^6, \\ 30\% \text{ by bus for cities } < 20 \times 10^3. \end{array}\right\} \qquad (7.6g)$$

(g) Peak-flow ratios.

Generally the minor flows in the peak are slightly higher than each of the equally divided off-peak in and out flows. The major direction of peak flow was 60–70% higher than the off-peak one directional flow.

(h) Duration of peak period

| | |
|---|---|
| Small cities | 45 minutes, |
| Medium | 75 minutes, |
| Large | 90 minutes. |

The independent variable $P$ was obtained from eight towns ranging from 18,000 to 8,000,000 inhabitants.

REGRESSION MODELS

Regression models have been exploited mainly in studies of the traffic generation of residential land-use zones. Ample data has been provided by the many home-interview studies that have been carried out in urban transportation surveys. Detailed patterns of home based trips by purpose or mode have been assembled and data on many of the independent variables has been obtained—vehicle ownership, economic and social status of family units, household structure, educational background and so on. Information on the land-use characteristics of the residential zones is also included—zoning, nature of industrial uses in the zone, density, distance from the city centre and so on.

One of the most useful studies of this kind was carried out on data obtained in the Washington D.C. Metropolis by Mertz and Hamner of the Bureau of Public Roads.[(63)] Another useful reference source is the ENO Publication *Highway Traffic Estimation.*[(64)] The factors commonly investigated are

vehicle ownership,
economic status,
population density,
distance from city centre.

Most investigations show that vehicle ownership has an overwhelming influence both on car trip and trips by all modes.[65] Further vehicle ownership is itself strongly correlated with household income—in the lower and median ranges particularly—and also with population density and distance from the centre.

The work of Mertz and Hamner yielded the following result for all trips—one way—per household:

$$T = 4{\cdot}33 + 3{\cdot}89V_1 - 0{\cdot}005V_2 - 0{\cdot}128V_3 - 0{\cdot}012V_4$$

where $V_1$ = vehicle ownership (cars/family unit),
$V_2$ = population density persons/acre,
$V_3$ = distance from central area—miles,
$V_4$ = economic index—family income in \$000's.

It is of some interest to note that one can establish from *a priori* argument a simplified relationship of the form

$$T = K(1+V) \tag{7.7}$$

where $K$ is approximately 4 and $V$ is the cars/family unit.

If $V = 0$ then an average family (say four or five persons) makes four outward trips per day. Such trips would include one for the bread-winner, one for the housewife and one each for say two school children —a rock-bottom minimum. It is not difficult to construct a schedule of ten to twelve outbound trips per day for a genuine two-car family or a single-car family where the car is available for the housewife.

However, an essential point to grasp is that whether the total number of trips be four or twelve there are in general only one or two work trips. Most of the problems of urban-transport planning are linked directly to getting the commuter to work either in his own car or by competitive public transport. The other three to ten trips per day are primarily local and there is no need to dwell on the fact that the traffic problems associated with intrazonal travel in residential areas are primarily related to parking.

Some comment should be offered also on the practice in the conventional transportation studies of dealing with the intermodal assignment issue in the traffic generation model. Certainly the early *O* and *D* studies were strongly oriented towards the solution of the car transport problem and as a result the traffic generation models were devised to give car trips directly. However, as work trips provide the primary peak-period traffic loading on the urban transport system it seems much more logical to obtain total work trips and carry out the assignment after the desire-line pattern has been obtained from the work trip ends.

## FUNCTIONAL MODELS

One of the principal defects of the empirical approach is that it does not permit the impedance of the transport system itself to be isolated. Any measured amount of traffic moving to or from a land-use zone is conditioned by the difficulty of making the journey. How then do we measure the intrinsic traffic generation capability of a land-use activity?

During recent years an approach utilizing the service-time concept of queueing theory has been applied to the study of the traffic generation of isolated, well-defined land uses such as supermarkets, service stations, drive-in food and drink facilities. Measurements have been made of the service-time characteristics of the functional and operational activities of these land uses. For example, in the case of a service station the distribution of the times taken to obtain petrol, oil, water, etc., has been determined and for parking lots the parking-time distribution. If the mean service time is $\bar{T}_s$ minutes then the turn-around rate at petrol pumps or parking bays is

$$\mu = 1/\bar{T}_s.$$

Further, if the traffic generation facility has $M$ service channels (four petrol pumps, 1000 parking spaces), then the maximum intensity of traffic generation of the land-use activity is

$$g_{\max} = \mu M. \tag{7.8}$$

Table 7.1 gives values of $\bar{T}_s$ and $\sigma_s$ for a variety of traffic generation facilities of the "drive-in" or neighbourhood kind.

It should be pointed out that the service-rate measure of traffic generation intensity gives a maximum generation rate. Such rates may occur during short peak periods. For busy periods of 1 to 4 hours duration it is necessary to introduce a utilization factor to ensure that waiting times in the queue do not become excessive. In activities of this kind a customer may be prepared to wait on the average about one service time in addition to his own time in service. In the single-channel situation this would require a utilization factor of 0·6–0·7. In cases where the number of service channels is large, say in parking areas, it is possible for the utilization factor to approach 1·0 without undue delay in the queue (ref. Table 3.1).

Some point should also be made of the fact that in many evaluation exercises it is convenient to base the analysis on a "worst case" situation—where the maximum generation rate is used to generate trip ends during a period of peak flow on the frontage road. If such a situation is tenable then no further evaluation is necessary.

The notion of functional behaviour may be applied to a large number of other traffic generation activities. In the case of theatres, sports arenas, restaurants, hotels, the trip ends may be evaluated directly in terms of the capacity of the respective activity. A theatre with seating for 2000 persons is capable of producing 2000 trip ends. The problem in this case is to determine the intensity of traffic generation. It is necessary to know the time period during which the trips are made.

TABLE 7.1. TRAFFIC GENERATION PARAMETERS FOR CERTAIN RETAIL ACTIVITIES IN THE SYDNEY REGION[19]

| Type of activity | Mean service time/channel $T_s$ (min) | Standard deviation $\sigma_s$ (min) | Max. generation rate $g$/channel (per hour) |
|---|---|---|---|
| Service station (single pump) | 3·0 | 2·1 | 20 cars |
| Supermarket cash-register channel | 1·2 | 0·9 | 49 persons |
| Pharmacy | 1·3 | 1·2 | 47 persons |
| Greengrocer | 3·5 | 2·9 | 17 persons |
| Butcher | 2·8 | 1·7 | 21 persons |
| Delicatessen | 1·6 | 1·0 | 37 persons |
| Dry cleaners | 1·7 | 0·8 | 35 persons |
| Hotel bottle dept. (drive in) | 1·9 | 1·1 | 33 cars |

Certain studies have been carried out at the Traffic Engineering School on the traffic generation of large office blocks and certain industrial establishments.[66,67] It has been found for most centres of congregation that it takes about an hour to reach the capacity when the land use is acting as an attractor (i.e. a destination). The emptying-out period is more variable. During loading up and discharge periods short-term peak overloads occur. Generalized conclusions from these studies were

(a) the average maximum intensity of traffic generation

$$\text{(filling)} = C/\text{hour},$$

$$\text{(emptying)} = \left.\begin{matrix}0\cdot 9\\ 1\cdot 2\end{matrix}\right\} C/\text{hour},$$

where $C$ is the capacity—patrons, employees, cars.

(b) the short-term peak overload factors were

filling 1·5,

emptying 2·0.

TRAFFIC GENERATION PARAMETERS

The ultimate rationale for a traffic generation model is to be found in the traffic generation parameter of a valid land use/transport interactance model. Consider the gravity model.

$$T_{od} = K\frac{L_o L_d}{Z(t_{od})}$$

$K$ is the traffic generation "constant". It is, however, a function of a number of socio economic, land-use and transport factors. We may write $K$ as

$$K = K(s, e, l, t, p)$$

where $s$ represents social factors such as educational background, family size,

$e$ economic factors such as income levels, taxation policies,

$l$ land-use considerations—density, zoning,

$t$ transport factors—vehicle ownership, public transport facilities,

$p$ the trip purpose.

The trip purpose would perhaps be better eliminated from any functional relationship by defining $K$ separately for each trip category, i.e.

$$K_p = K(s, e, l, t).$$

The nature of this function might well be explored by regression analysis. However, there are three important situations that need to be considered in pursuing the discussion on the traffic generation parameter, viz.

(a) where the generation is constrained by the origin trip ends,

(b) where it is constrained by a fixed amount of travel from the origin,

(c) unconstrained.

### *Fixed Number of Trips*

If the pattern of trip ends is given then it may be inferred that the traffic generation has been determined. This leads to the elimination of $K$. The trips from an origin zone $o$ must equal the trips to all destination zones

$$d = 1, 2, \dots, D,$$

therefore

$$L_o = T_{o1} + T_{o2} \cdots T_{oD}$$

but

$$T_{o1} = K\frac{L_o L_1}{Z(t_{o1})}, \qquad T_{o2} = K\frac{L_o L_2}{Z(t_{o2})}, \quad \text{etc.}$$

therefore

$$L_o = KL_o \sum_{d=1}^{D} \frac{L_d}{Z(t_{od})}$$

and

$$K = 1 \Big/ \sum_{d=1}^{D} \frac{L_d}{Z(t_{od})} \tag{7.9}$$

whence the well-known distribution formula for a given trip end pattern, i.e.

$$T_{od} = L_o \frac{T_d/Z_{od}(t)}{\sum_d T_d/Z_{od}(t)} \qquad \begin{array}{l} o = 1, 2 \dots O, \\ d = 1, 2 \dots D. \end{array}$$

This form of the land-use transport equation is particularly appropriate to the generation and distribution of work trips where there is little difficulty in relating the total number of trips to the population and jobs of the origin and destination zones.

### *Fixed Amount of Travel*

In an interesting paper[(3)] on the factors affecting the amount of travel in the U.K., Tanner has presented data to support the proposition that individuals spent a fixed proportion of their income on travel. Thus if land-use or transport changes resulted in a reduction in the cost of travel more trips would be made. It is not difficult in view of the generally linear relationship between the cost of a trip and its length or time, to express Tanner's proposition in terms of person miles or person hours, i.e. that the average person travels a fixed number of miles or hours during a given real time period.

In his paper Tanner considers trips from $d$ to $o$ as well as those from $o$ to $d$. The question may well be asked as to whether this is correct. We have indicated earlier that trips have a vector character at least when they still represent a demand. It is appropriate to raise here the confusion that has been introduced into land-use and transport planning by the widespread American use of the Average Daily Traffic concept. The A.D.T. measure has its origin in rural road planning where a desire line is generally synonymous with the traffic flow. Further there is little need to differentiate sharply between the different trip purposes in intercity travel nor is there a need to distinguish between the directional flows on a two-lane, two-way rural road because of its somewhat special property of providing a more or less constant practical capacity irrespective of the directional flow ratio.

The situation in the urban scene is quite different in all these respects and it is important to distinguish direction and purpose for each desire line.

Suppose the average time spent by an individual in a given real time period (say a day) in travelling about is $H$. Suppose, too, that the times taken to travel between an origin zone $o$ and a series of destinations $d = 1, 2, \ldots, D$ are $t_{o1}, t_{o2}, \ldots, t_{oD}$. Then the total time spent in travel by the origin land users is $L_o \times H$ and the time spent in transit between $o$ and the destination is

$$T_{o1} \cdot t_{o1} + T_{o2} \cdot t_{o2} \ldots + T_{oD} \cdot t_{oD}.$$

Therefore

$$L_o \times H = KL_o \left[ \frac{L_1 \times t_{o1}}{Z(t_{o1})} + \frac{L_2 \times t_{o2}}{Z(t_{o2})} \cdots \frac{L_D \times t_{oD}}{Z(t_{oD})} \right]$$

whence

$$K = H \Bigg/ \sum_{d=1}^{D} \frac{L_d t_{od}}{Z(t_{od})} \tag{7.10}$$

and

$$T_{od} = HL_o \frac{L_d / Z(t_{od})}{\sum_{d=1}^{D} \frac{L_d \cdot t_{do}}{Z(t_{do})}}.$$

***Unconstrained Trip Ends***

This situation applies where the number of trips leaving a given origin zone is small compared with the total number of origin trip ends.

In this case the traffic generation parameter $K$ may be regarded as a "universal" constant in any given context of socio economic and other relevant circumstances.

This notion of $K$ is particularly applicable to intercity travel. A number of investigations relating to intercity air travel, travel between towns and provincial cities and special situations relating to recreational travel give relatively consistent "constants" for $K$. In an unpublished pilot study carried out in 1963, relating to the traffic and revenue prospects of a toll road between Sydney and Newcastle, useful values of the constant were obtained. Sydney ($2{\cdot}5 \times 10^6$ population) and Newcastle ($2{\cdot}0 \times 10^5$) are about 70 miles apart on the central coast of New South Wales. In between there are five important lakeside and coastal recreational centres. Values of $K$ were determined for different trip categories. With travel time in minutes, approximate values were,

| | |
|---|---|
| Recreational | $K = 0{\cdot}05$ |
| Business work and shopping | $K = 0{\cdot}5$ |
| Private social | $K = 0{\cdot}005$ |

Studies relating to travel between rural towns in Victoria and Queensland give consistent $K$ values.

If a "constant" $K$ can be found for rural travel it is possible to devise some simple formulae for traffic growth and induced or generated traffic. Suppose

$$T_{od} = K\frac{L_o L_d}{t_{od}^2}.$$

Then for the general growth trend $\delta T_{od}$ based on growth of the land-use potential

$$T + \delta T_{od} = \frac{K}{t_{od}^2}(L_o + \delta L_o)(L_d + \delta L_d)$$

$$= \frac{K}{t_{od}^2}(L_o \,.\, L_d + L_o \delta L_d + L_d \delta L_o + \delta L_o \,.\, \delta L_d).$$

Ignoring second-order effects

$$\frac{\delta T_{od}}{T_{od}} = \frac{\delta L_o}{L_o} + \frac{\delta L_d}{L_d} \tag{7.11}$$

a result except for a factor of 2 similar indeed to the once popular average factor method.

If we wish instead to consider the traffic induced on account of the transport improvements alone the induced traffic $\delta T_{od}$ is given by

$$\delta T_{od} = \left[\frac{(t_{od}^2)\text{ after}}{(t_{od}^2)\text{ before}} - 1\right] T_{od}. \tag{7.12}$$

In the rural context where it is appropriate to consider car trips separately it is helpful to use vehicle populations of the land-use zones as the measure of land-use potential.

## SOME OBSERVATIONS AND EXAMPLES

Although the traffic engineer has made a notable contribution to the early practice of assessing the total amount of traffic that is "given-off" from various types of land use by virtue of his massive transportation studies and extensive cordon counts, the grossness of the approach has clouded the true significance of many of the basic traffic-generation parameters. The darkest cloud hangs over the daily traffic concept, for with its inclusion of all types of trips the number of independent variables that influence this measure are unduly numerous, vague and themselves hard to predict. Also regression analyses based on a space cross-section of independent variables has dubious merit in determining their influence in extended real time.

The principal objectives of the foregoing discussion have been

(a) to distinguish between the main logistical variables, i.e. land-use potential and transport impedance, and the inherent traffic generation propensity of the land-use activity in a functional sense;

(b) to emphasize the importance of the time and space dependence of the intensity of traffic generation.

The significance of these important aspects will be best illustrated with some specific examples.

### *Long-term, Large-scale Urban Planning*

In this context one of the prime objectives of the land-use transport planner is to study the system under conditions of peak loading of the trunk transport network. Although it is not possible to phase the different trip categories into exact time slots there is little doubt that the rationale of traffic estimation is well served by the assumption that work trips together with some school trips dominate the week-day traffic pattern. Further, work and school trips represent a relatively inelastic demand and are highly predictable, even in the long term. All that is required of the traffic-generation model is an estimate of the total number. The land-use and desire-line models can then distribute their origins and destinations amongst the various zones on the land-use plan.

If we turn now to ANZAC City it will be clear that the work-trip traffic generation is completely determined by the present number of jobs and the estimated increase in them (i.e. 22,500 + 14,500 = 37,000). This total is, of course, usefully constrained by the present distribution of job locations and residential origins of the workers. Appropriate basic data is given in Table 1.2.

The origins of the work trips could in general be realistically specified from available and authentic data. In the case of our hypothetical ANZAC City the distribution of residential origins of the work force is assumed to be proportional to the total population of the residential zones. Here the proportionality constant is 22,500/67,000, i.e. 0·34.

At this point we must raise the important question of trip stratification. In this particular example the destinations are in fact stratified into white-collar and black-shirt categories. It is necessary therefore to differentiate the origins. This, too, has been done by simple proportioning, i.e. in each residential zone the worker origins have been divided into two groups in the ratio of 14,500:8000. Thus there are 1·8 industrial workers for each white-collar resident. Here again realism would be increased by using sample data in dealing with an actual system. However, the important point to make is that the sub-totals of the residential origins over all zones should match the number of jobs available in these categories.

It should be pointed out that considerable improvement in forecasting and much better understanding of the peak-hour phenomenon can be gained from finer stratification—male, female, executive, retail, skilled artisans and so on. Salary classification is also useful.

The future situation calls only for an estimate of the totals of new jobs in the various categories. These may be arbitrarily allocated to the land-use zones and a desire-line pattern then obtained (refer to Chapter 8) or used to "scale" a combined land-use and desire-line model.

If, however, the assignment phase is to be incorporated it is necessary to convert the demand into an intensity measure. To do this it is necessary to know the time period over which the demand is valid. In any large system it is an extremely difficult task to determine the space distribution of traffic generation intensities and little work on the problem has as yet been attempted. However, on the cordon of the major destination zones there is a good deal of evidence to support the contention that the concentration of a set of trip desires is spread over a period of about 1 hour. The findings of the R.R.L. Study[62] provide firm grounds for this. Where the intensity of demand is well below the saturation level of the transport facilities the period required to "focus" a concentration on a given starting time seems to start about 60 minutes beforehand, but there is generally a 10-minute peak overload period much closer to the concentration time. During this period intensities rise to about twice the average. In the case of ANZAC City the peak demand rate is assumed to occur uniformly over a 1 hour period. In large cities account should be taken of situations where there is well-defined staggering of major work trip strata.

Table 8.1 establishes the traffic generation characteristics for work trips in ANZAC City.

### *Educational and School Trips*

It may be taken generally that trips for higher education—universities, technical institutes, colleges—are included in the work-trip category. Other school trips are essentially local and may be ignored in the analysis of the trunk transport system. Such trips are of course of great interest to the land-use and transport planner from the point of view of exact location, parking, access details and so on. For the treatment of such aspects traffic generation estimates are needed. As in the case of work trips this is simply based on total student numbers in each school zone. In the case of our hypothetical city it has been assumed that there are no institutes of higher education and all educational trips are intrazonal.

### *Shopping, Recreational and Social Trips*

In general these classes of trips do not occur in the week-day rush hour. However, in certain instances they create important peak demands at other times—football finals, holiday week ends, for example. They are primarily home-based trips and their distribution pattern is firmly shaped by the land-use potential of the alternative destinations and the transport impedance between origin and destination. There is, however, a considerable "elasticity" in the demand for these kinds of trip. Tanner's study[(3)] lends support to the constant amount of travel theory. Whilst there is little authentic data on the ration of travel ($/week, hours/month) for different categories of trips and various classes of travellers it is common to assume that in high standard-of-living communities individuals spend on the average some 500 hours/year travelling. If travel to work represents, say, 50% of this, then the weekly travel parcel for non-work trips would be of the order of 5 hours. Shopping trips may reasonably be assumed to utilize about 3 hours/week of shoppers' time. It will be instructive for the student to make some calculations of shopping-trip generation rates for various travel circumstances—2, 3, 4 hours/week; travel by public transport to central area, walking to neighbourhood centres and car travel to regional centres. Such an exercise would guide the student into some very profitable research on the time spent in travel for various trip kinds.

### *Traffic Generation Based on Functional Activity of Zones*

The functional model approach is very useful for determining the traffic-generation characteristics of certain shopping and recreational activities—particularly in car-oriented situations, e.g. supermarkets, regional centres, drive-in restaurants, bars, clubs, bottle departments and so on. In these cases the generation characteristics are simply related to the parking capacity and the turn-around rate. The utility of this kind of approach is best demonstrated by an example.

A small supermarket has parking space for 100 cars. On Friday—the busiest trading day—it is full by 10 a.m. and remains so until about 1 p.m. when the accumulation of cars drops to fifty between 12.45 p.m. and 1.10 p.m. It remains at that level until 2 p.m. By 2.30 p.m. it is full again; and at 4 p.m. shoppers gradually disperse. By 5.30 p.m. it is empty. A study of the distribution of parking times shows that it is exponential with a mean of 40 minutes. Find

(a) the traffic generation intensities throughout the day;

(b) the total trip ends for the day's trading;

(c) the maximum traffic flow caused by the centre.

The following simplified analysis assumes that during the main filling-up and discharge periods there is negligible turn-around activity. The accumulation pattern is

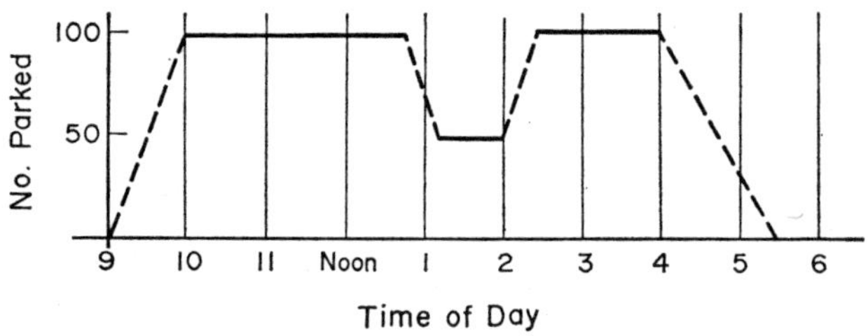


The time pattern of the intensity of traffic generation with the parking lot acting as a destination and an origin is shown on facing page.

From 9 a.m. to 10 a.m.: 100 vehicles arrive in 1 hour $\therefore g_d = 100/\text{hr}$;

10 a.m. to 12.45 p.m.: $g_d = g_o = M \cdot \mu = 100 \times 60/40 = 150/\text{hr}$;

12.45 p.m. to 1.10 p.m.: 50 vehicles depart in 25 minutes, $\therefore g_o = 120/\text{hr}$;

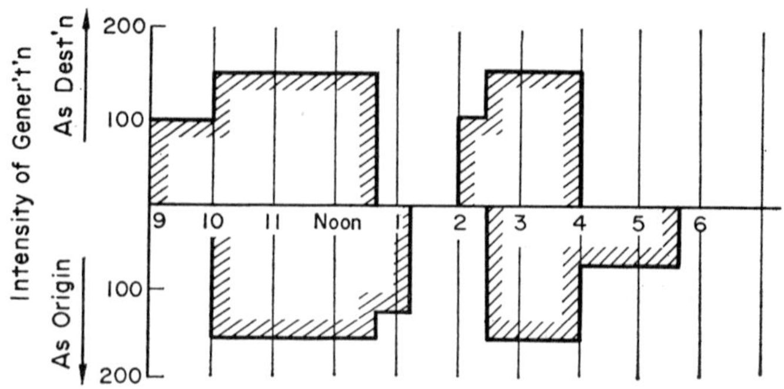


1.10 p.m. to 2.0 p.m.: no activity $\therefore g = 0$;

2 p.m. to 2.30 p.m.: 50 vehicles arrive $\therefore g_d = 100$/hr;

2.30 p.m. to 4 p.m.: $g_d = g_o = M \cdot \mu = 100 \times 60/40 = 150$/hr;

4 p.m. to 5.30 p.m.: 100 vehicles depart $\therefore g_d = 67$/hr.

Total trip ends

as destination: $100 \times 1 + 150 \times 2{\cdot}75 + 100 \times 0{\cdot}5 + 150 \times 1{\cdot}5 = 787$,

as origin: $150 \times 2{\cdot}75 + 120 \times 0{\cdot}42 + 150 \times 1{\cdot}5 + 67 \times 1{\cdot}5 = 787$.

The greatest rate of flow associated with the supermarket arises from max. $(g_d + g_o)$ at a given time point or interval. Here

$$q_{\max} = 150 + 150 = 300 \text{ veh./hr}$$

during the a.m. and p.m. main trading hours.

CHAPTER 8

# Desire Lines

## INTRODUCTION

A desire line is a directed line from an origin to a destination representing the demand for communication between them. It is usually drawn of varying width to approximately scale the level of the demand in units of trips/unit time, i.e. per hour, per day, per year. The trip may be measured in passenger, ton, vehicle, units. In the case of telephone traffic the unit becomes an event, i.e. a call. A desire line is not synonymous with traffic flow except perhaps in the case of rural traffic, where a series of towns are directly connected together. A desire-line pattern comprises a set of independent desire lines between a set of origins and destinations. There are $m \times n$ possible desire lines between $m$ origins and $n$ destinations.

A desire-line pattern may be presented graphically or numerically in a two-dimensional array. In such format the independent variables are the space coordinates. But real time is also an essential variable and up until recently the time variable has been to a large extent ignored by the simple expedient of using the average daily demand as the measure of the desire lines. Generally the average daily desire is a weekday desire and few transport studies give specific consideration to week-end desire-line patterns.

The dominance in much of urban transport planning of the average daily traffic (A.D.T.) measure has its origins in rural road planning where measured traffic volumes on main intercity and intertown roads were for all practical purposes desire-line measures. In the urban scene traffic patterns have a well-defined daily cycle, at least for weekdays, and the characteristics of this cycle are strongly related to the purpose of the trip—work, shipping, school, recreational.

An urban desire-line pattern should thus be defined in one of two ways:

(a) by hour of day,
(b) by trip purpose.

The most adaptive of these alternatives is the latter, especially as the work trips desire-line pattern establishes the principal criteria for planning and design purposes. We shall see later on that in the context of long-term planning the future distribution of employment and residential zones is more predictable and perhaps more controllable than other zonal functions. The specification of desire lines in terms of the A.D.T. measure makes them non-directed and this results in certain difficulties so far as capacities and travel times are concerned, especially if more generalized network methods are employed in subsequent analysis.

It has been stated above that the desire-line measure may be passenger, freight or vehicular trips. If the latter, then some inference must be drawn concerning the assignment of the basic passenger or freight desire between modes. The assignment problem will be taken up in Chapter 9. At this point it is sufficient to say that assignment between modes may be carried out before or after the desire-line pattern has been determined. If it is carried out before, then a further replication of patterns is necessary, i.e. car, rail, bus, walking, ferry.

It should be clear then to the student that an enormous amount of data is implicit in the definition of the desire-line pattern of an urban area and a great deal of thought needs to be given to its assemblage if it is to be meaningfully exploited. The student should ponder thoughtfully the consequences of dividing a major metropolitan area into 1000 land-use zones—a tempting prospect when one considers the facility of the computer and the diversity of land-use activity of a square mile of urban land, but a frightening one when it is realized that 1000 zones give rise to one million independent desire lines. One knows, of course, that the pattern of arrivals at a regional shopping centre, or the proportion of "through" trips in Main Street, require specification, just as does the commuter demand in a major transport corridor. The important point, however, is that they should be studied as separate problems even though there may be some interrelation between them.

The development of techniques to "partition" a large-scale problem into a set of smaller problems is one of the essential requirements of the systems approach.

## DETERMINATION OF DESIRE-LINE PATTERNS

Notwithstanding the many unresolved complexities of properly describing a desire-line pattern, the determination of such patterns is without doubt the aspect of the transport planning methodology that has achieved some degree of maturity in a remarkably few years. Little more than a decade ago the approach was dominated by the rural road planner and for the reasons mentioned above the determination of future patterns was based on historical growth prediction of an existing flow pattern. It was soon found to be inadequate for the urban scene and the concept of some interaction between land use and transport was intuitively introduced in the average factor and Fratar methods.

In both cases an existing desire-line pattern, obtained from surveys, was extrapolated by a factor

$$\tfrac{1}{2}(F_o + F_d)$$

where $F_o$ and $F_d$ were the growth factors of the origin and destination land uses. These growth factors involved the traffic-generation phase of the transport planning process for they represent the ratios of future to present trip ends originating or terminating in the origin and destination zones respectively. In some cases growth factors were based on present and future zonal population ratios and in others these ratios were multiplied by a further factor which took account of the increase in vehicle ownership.

In both the average factor and Fratar methods it was necessary to make successive iterations in order to ensure that the sum of the future desire lines at origin and destination converges towards the estimated future trip ends. There is little point in regarding these methods as having other than historical interest nowadays, partly because they take no account of the effect of the transport factor and partly because of the power and versatility of the gravity model and certain other models.

## THE GRAVITY MODEL AS A DESIRE-LINE ESTIMATOR

There is an extensive literature[68–74] on the application of the gravity model to desire-line estimation and its use in this role is supported by a variety of computer packages. In some respects it has been over-exploited. The relative simplicity of its use has encouraged the analyst to over-subdivide the study area and as a result 100's, even 1000's, of land-use zones are defined. The enthusiasm to calibrate it with current data has tended to make it something more akin to a regression-line model than a creative planning tool.

We have seen already that a gravity-type formula permits the traffic demand between land-use zones to be defined in terms of the key concepts of land-use potential and transport impedance. But the traffic yield from the formula requires the constant of proportionality $K$ to define and measure the traffic generation of the system.

It is common in using the gravity model for desire-line patterns to assess the traffic generation independently by defining the trip ends. The proportionality constant may thus be eliminated and the distribution formula is

$$T_{od} = L_o \frac{\dfrac{L_d}{Z(t_{od})}}{\sum\limits_{d=1}^{D} L_d/Z(t_{od})} \qquad o = 1, 2, \ldots, O.$$

Although the traffic generation constant has been eliminated it is common practice to calibrate the above form of the distribution model. This is done partly by determining from given desire-line data an empirical form of $Z(t_{od})$ and in addition a special attraction factor $A_d$ used to modify the land-use potential $L_d$. However, there is an increasing awareness of the fact that calibration in terms of existing travel practices is tantamount to converting this powerful planning formula into a regression equation and in consequence its value as a creative planning tool is diminished. A major reason for the introduction of the special attraction factors arises from the practice of using the formula to synthesize average daily desire-line patterns. If instead distribution of the main trip purpose categories is carried out separately the need to determine special attraction factors is of much less importance—particularly for work trips. And as work trips are in the main responsible for the peak demands for transport in urban areas there seems to be good reason for focusing attention on this class of trip making. School, business and shopping trips are also relatively insensitive to special factors. By contrast recreational and social trips are often strongly influenced by special land-use characteristics.

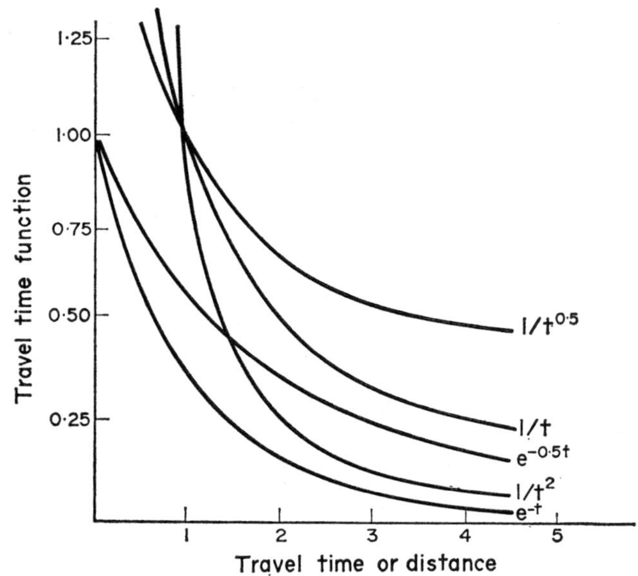


FIG. 8.1. The nature of the travel-time function.

The travel-time function has been the subject of a good deal of study. If we define

$$Y(t_{od}) = \frac{1}{Z(t_{od})}$$

then Fig. 8.1 shows a number of commonly employed $Y$-functions. Although these functions are significantly different as plotted it is relevant to observe that for values of the independent variable greater than 2 they are sensibly parallel to one another. In the gravity model formula the effect of a scaling change in the value of $Y(t_{od})$ or $Z(t_{od})$

does not alter the value of $T_{od}$ because these functions have the same dimension in both numerator and denominator. It is not surprising therefore that the gravity-model-distribution formula is relatively insensitive to the travel-time function—for the longer trips anyway.

Even for values of $t$ less than 2 there is a surprising degree of parallelism between $e^{-t}$ and $1/t$ and between $e^{-0.5t}$ and $1/t^{0.5}$. On the other hand, it is clear that in this lower range travel time is a powerful discriminator of trips.

One other point of interest is the observation of Tanner[(3)] that a simple inverse function of $t$ produces singularities at the origin and at infinity in certain theoretical situations. Tanner proves that the travel-time exponent must be less than 3 for traffic in towns (if very small interzonal distances are to be considered). On the other hand, it needs to be greater than 3 if intercity or large urban distances are to be properly represented. Although Tanner points out that the commonly used values of between 2 and 3 flout these conditions the fact of the matter is that values in this range come reasonably close to meeting the theoretical constraint. In practice, however, the difficulty is more apparent than real for $t$ seldom approaches either zero or infinity. However, the theoretical difficulties are overcome if the exponential versions of the travel-time function are used.

## An Example

Because we shall have a good deal to say later on concerning the comparative merit of the Gravity Formula and Linear Programming for desire-line determination we shall consider here the problem already used in Chapter 4 to illustrate the solution of the Linear Programming Transportation Problem. The situation is illustrated in Fig. 4.9. The essential data may be summarized using the Land-use Matrix (see next page) described in Fig. 1.2, Chapter 1.

The travel times shown are the minimum time paths of the transport network of Fig. 4.9.

The gravity formula may be written

$$T_{od} = R_o \frac{E_d/t_{od}^n}{\sum_d E_d/t_{od}^n}.$$

| o \ d | 1 | 2 | R's |
|---|---|---|---|
| 1 | $T_{11}$ / 11 | $T_{12}$ / 22 | 3000 |
| 2 | $T_{21}$ / 16 | $T_{22}$ / 18 | 3000 |
| 3 | $T_{31}$ / 21 | $T_{32}$ / 10 | 3000 |
| E's | 4000 | 5000 | |

We shall consider solutions for $n = 0, 2, 4$.

The student will have little difficulty in applying this formula as a sample calculation will show.

For $n = 0$

$$T_{11} = R_1 \frac{E_1}{E_1+E_2} = 3000 \frac{4000}{4000+5000}$$

$$= 1333,$$

$$T_{12} = R_2 \frac{E_2}{E_1+E_2} = 1667.$$

Continuing thus we obtain

| $o$ \ $d$ | 1 | 2 | $\sum_d T_{od}$ |
|---|---|---|---|
| 1 | 1330 | 1667 | 3000 |
| 2 | 1330 | 1667 | 3000 |
| 3 | 1330 | 1667 | 3000 |
| $\sum_o T_{od}$ | 4000 | 5000 | |

Total trip minutes = 147,190.

It should be noted that in this solution the trip-end constraints are met exactly. In fact the above solution is a feasible (though not a basic) solution of the L.P. problem of Chapter 4.

For $n = 2$.

Here

$$T_{11} = R_1 \frac{E_1/t_{11}^2}{E_1/t_{11}^2 + E_2/t_{12}^2}$$

$$= 3000 \frac{4000/11^2}{4000/11^2 + 5000/22^2} = 2280.$$

Clearly

$$T_{12} = 3000 - 2280 = 720.$$

Proceeding similarly for $T_{21}$, $T_{22}$, $T_{31}$ and $T_{32}$ we obtain

| $o \backslash d$ | 1 | 2 | $\sum_d T_{od}$ |
|---|---|---|---|
| 1 | 2280 | 720 | 3000 |
| 2 | 1510 | 1490 | 3000 |
| 3 | 460 | 2540 | 3000 |
| $\sum_o T_{od}$ | 4260 | 4740 | |

Total trip minutes = 126,610.

It will be seen that whilst a trip-end balance exists at the origin end, the job end summations do not match the respective $E$'s.

The situation may be improved by iteration based on adjustments to the $E$'s, viz.

$$E_d' = \frac{E_d}{\sum_o T_{od}} \cdot E_d \qquad \text{for} \quad d = 1, 2.$$

A new set of $T_{od}$ may now be obtained, viz.

| $o \backslash d$ | 1 | 2 | $\sum_d T_{od}'$ |
|---|---|---|---|
| 1 | 2220 | 780 | 3000 |
| 2 | 1425 | 1575 | 3000 |
| 3 | 418 | 2582 | 3000 |
| $\sum_o T_{od}'$ | 4063 | 4937 | |

Total trip minutes = 127,070.

The trip-end balance is now sufficiently close. However, if further iterations are necessary it should be noted that

$$E_d'' = \frac{E_d'}{\sum_o T_{od}'} \cdot E_d \quad NOT \, . \frac{E_o'^2}{\sum_o T_{od}'}$$

It is pertinent to raise the question whether there is real point in seeking a trip end balance in this manner. Why not adjust the job-trip ends on completion of the initial solution? The duality between the desire-line and land-use allocation problems would support this proposal. Further, when a desire-line pattern is constructed for a future estimate of trip ends the accuracy of such an estimate is not likely to be any more accurate than the initial $\sum_o T_{od}$ values.

Further it will be noted that the first iteration gives a smaller value for the total trip minutes of travel.

For $n = 4$. The results for this case are as shown below.

| $o \backslash d$ | 1 | 2 | $\sum_d T_{od}$ |
|---|---|---|---|
| 1 | 2602 | 398 | 3000 |
| 2 | 1195 | 1805 | 3000 |
| 3 | 62 | 2938 | 3000 |
| $\sum_o T_{od}$ | 3559 | 5141 | |

Total trip minutes = 119,760.

## OPPORTUNITY MODELS FOR DESIRE-LINE DISTRIBUTION

In 1940 Stouffer[75] proposed a theory of intervening opportunities and Schneider exploited the concept in developing an intervening opportunities model for the Chicago Area Transportation Study.

The opportunities models do not introduce distance or time explicitly but take account of the effect of spatial separation of zones by ranking destinations (or origins) in time or distance order from a given origin (or destination). More specifically the Schneider model[76] rests on the proposition that the probability that a trip will terminate in a given zone is proportional to the land-use potential of that zone times the probability that a destination closer to the origin zone has not been selected.

If $P(L)$ is the probability that a trip will find a destination amongst a distribution of land uses $L$. Then the probability that a trip will terminate in an elementary land-use region $dL$ on the fringe of the distribution is

$$dP(L) = K\,.\,\{1-P(L)\}\,dL.$$

Solution of this equation with the initial conditions $P = 0$ when $L = 0$ gives

$$P(L) = 1-e^{-KL}.$$

Now the number of trips from the origin to zone $d$ is

$$T_{od} = L_o\{P(\sum_{d=1}^{D} L_d)-P(\sum_{d=1}^{D-1} L_d)\}$$

or

$$T_{od} = L_o\left\{\exp\left(-K\sum_{d=1}^{D-1} L_d\right)-\exp\left(-K\sum_{d=1}^{D} L_d\right)\right\}.$$

The parameter $K$ may be thought of as a siting function. In practice it is difficult to determine—it is analogous to the traffic-generation parameter of the gravity model.

However, in the simple distribution role Tomazinis[77] has proposed an approach which eliminates the need to obtain it explicitly by using the trip ends as measures of the land-use potential. This is analogous also to the gravity model approach to the distribution problem with a predetermined trip-end pattern.

Tomazinis' method may be stated

$$P_{od} = \frac{E_d}{\sum_d E_d} \Big/ \sum \left( \frac{E_d}{\sum_d E_d} \right)$$

with the destination zones $E_d$ ranked in order of their time or distance from the origin $o$. The desire line $T_{od}$ is then

$$T_{od} = R_o P_{od}.$$

The application of this method may be simply illustrated by returning to our $3 \times 2$ work-trip problem.

For origin zone $R_1 = 3000$

| $d$ | $E_d$ | $\Sigma E_d$ | $E_d/\Sigma E_d$ | $P_{1d}$ | $T_{1d} = R_1 P_{1d}$ |
|---|---|---|---|---|---|
| 1 | 4000 | 4000 | 1·00 | 0·65 | 1950 |
| 2 | 5000 | 9000 | 0·55 | 0·35 | 1050 |
| | $\Sigma(E_d/\Sigma E_d)$ | | 1·55 | | |

If we proceed similarly for $R_2$ and $R_3$ we obtain the travel desire matrix

| $o$ \ $d$ | 1 | 2 | $\sum_o T_{od}$ |
|---|---|---|---|
| 1 | 1950 | 1050 | 3000 |
| 2 | 1950 | 1050 | 3000 |
| 3 | 900 | 2100 | 3000 |
| $\sum_d T_{od}$ | 4800 | 4200 | |

Total trip minutes = 134,550.

It will be noted that the trip-end balance for the destinations is rather badly out. This illustrative example provides a rather extreme situation for a method such as this. Improvements can be made by employing an iterative procedure similar to that explained for the gravity model. It will be noted that the total trip minutes is greater than that obtained from the application of the gravity formula.

LINEAR PROGRAMMING VERSUS DEDUCTIVE MODELS

In the L.P. solution of this problem given in Chapter 4 the following points may be noted:

(a) The trip minutes are less than any of the solutions given above. This of course is to be expected.

(b) The trip end constraints are met exactly.

(c) Only 4 desire lines have non-zero values. In general the L.P. solution gives $(m+n-1)$ non-zero values whereas the gravity and other models yield a value for all $m \times n$ desire lines. In this case the difference between $m+n = 6$ and $(m+n-1) = 4$ is not marked. However, for a $10 \times 10$ system would evaluate only 19 of the 100 possible desire lines.

This latter situation would seem on first consideration, to seriously handicap the L.P. method. However, even in this simple example the L.P. values are clearly the major ones and from the point of view of long-term planning there are many advantages in a solution that highlights what should be done rather than one that perpetuates existing behaviour.

The above observation should, of course, be qualified by adding that many additional constraints (e.g. capacities, travel times) may be added and an extra desire line comes to light with each addition. Further, the L.P. desire-line pattern is diversified if the trip ends are stratified. And this is something that is most desirable in any case. The generally good results of the gravity model are due in large measure to its lack of discrimination of trip categories. Its use with average daily traffic desires seems to produce fortuitously good results.

The most important consequence of the linear programming solution of the desire-line problem arises from the fact that it provides perhaps the best validation of the gravity model approach. It will be noted in the example dealt with above and further confirmed in the results for ANZAC City that the total trip minutes for the various solutions are not appreciably greater than the L.P. optimum. In other words, the gravity model looks for "good" solutions. In fact one might well regard the gravity formula as a device for obtaining an iterative solution to the

L.P. problem. It will have been seen, too, that as the exponent of $t$ increases the gravity model results approach more closely those obtained by programming.

### *Desire-line Studies in the Sydney Metropolitan Region*

A highly reputable sample of 80,000 work trips was made available in 1965 by the Commonwealth Bureau of Census and Statistics.(78) The sample was taken from a 100% data base obtained in the 1961 National Census. The Sydney Metropolitan Area is divided into forty municipal areas and shires. Bell(79) has used this data to test a variety of distribution models and Connors(80) aggregated the forty municipal zones into twelve macro zones to test and compare a L.P. solution.

Bell tested

(a) the opportunity model of Tomazinis;

(b) gravity model with a simple inverse distance function with $n = 2$;

(c) gravity model using an exponential function $e^{-0{\cdot}08d}$;

(d) a multiple regression model—

$$y = a+bL_1+cL_2+dL_3+eL_4+fL_5+gL_6+hL_7+iL_8$$

where $L_1$ = direct distance between zones,

$L_2$ = attraction probability as obtained from intervening, opportunities model,

$L_3$ = resident population of zone,

$L_4$ = percentage of residents employed in manufacturing,

$L_5$ = distance of each zone from Central Area,

$L_6$ = average number of persons/room in each zone,

$L_7$ = dummy variable indicating whether zone is north or south of harbour,

$L_8$ = dummy variable indicating whether zone is served by rail transport.

The validity of the various models was tested by forming a trip-length frequency distribution of actual and predicted trips and computing the r.m.s. error, i.e.

$$\text{r.m.s. error} = \sqrt{\left(\frac{T_{od} - T^*_{od}}{n}\right)}$$

$T_{od}$ and $T^*_{od}$ are the actual and estimated trips, respectively, and $n$ the number of trips.

All models gave a realistic synthesis, the most accurate being the regression model, then in order the inverse square, exponential gravity models and the intervening opportunity model.

In the comparison between the L.P. and gravity model for the Sydney region the reason for consolidating the forty local government zones into twelve super zones was necessary in order to permit the stratification of the work force into white- and blue-collar jobs and male and female components.

The results showed

(a) good agreement between the L.P. and gravity model (with $n = 2$) for intra-zonal trips;
(b) a generally similar interzonal desire-line pattern;
(c) a total travel activity of 367,000 trip hours as compared with 421,000 actual and 477,000 by the gravity model.

## DESIRE-LINE PATTERN FOR ANZAC CITY

The application of the gravity model to ANZAC City is left as a relatively straightforward exercise for the student. The results are tabulated in Table 8.3. The following points should be noted:

(a) travel times are those given in Table 1.3;
(b) both the jobs and residential locations of the work force are stratified.

The jobs are stratified into commercial and retail (all located in zone 8) and industrial jobs are located in zones 1, 6 and a small part of zone 8. In order to compare the results with the desire-line pattern that emerges

TABLE 8.1. STRATIFICATION OF ORIGIN AND DESTINATION

| Zone no. | Residential origins | | Destinations | |
|---|---|---|---|---|
| | Indust. | R. and C. | Indust. | R. and C. |
| 1 | | | 9200 | |
| 2 | 2200 | 1500 | | |
| 3 | 4000 | 2500 | | |
| 4 | 1500 | 1000 | | |
| 5 | 3700 | 2800 | | |
| 6 | | | 11,600 | |
| 7 | 1300 | 800 | | |
| 8 | | | 1900 | 14,300 |
| 9 | 2000 | 1200 | | |
| 10 | 3500 | 2400 | | |
| 11 | 1000 | 1400 | | |
| 12 | 3500 | 1500 | | |
| Total | 22,700 | 14,300 | 22,700 | 14,300 |

Table 8.2 provides a simple format for the calculations.

TABLE 8.2.

| Origin zone no. | Origin (1) | Impedance function $o$ to $d$ | | | $\frac{E_1}{t_{o1}^2}$ (2) | $\frac{E_6}{t_{o6}^2}$ (3) | $\frac{E_8}{t_{o8}^2}$ (4) | $\Sigma \frac{E_d}{t_{od}^2}$ (5) |
|---|---|---|---|---|---|---|---|---|
| | | $t_{o1}^2$ | $t_{o6}^2$ | $t_{o8}^2$ | | | | |
| 2 | 2200 | 18·1 | 176·0 | 110·0 | 0·51 | 0·07 | 0·02 | 0·60 |
| 3 | 4000 | 20·2 | 90·5 | 100·0 | 0·46 | 0·13 | 0·02 | 0·61 |
| 4 | 1500 | 72·5 | 182·0 | 121·0 | 0·13 | 0·06 | 0·02 | 0·21 |
| 5 | 3700 | 42·5 | 56·2 | 64·0 | 0·22 | 0·21 | 0·03 | 0·46 |
| 7 | 1300 | 66·0 | 244·0 | 112·0 | 0·14 | 0·05 | 0·02 | 0·21 |
| 9 | 2000 | 132·0 | 49·0 | 64.0 | 0·07 | 0·24 | 0·03 | 0·34 |
| 10 | 3500 | 182·0 | 49·0 | 100·0 | 0·05 | 0·24 | 0·02 | 0·30 |
| 11 | 1000 | 92·0 | 293·0 | 150·0 | 0·10 | 0·04 | 0·01 | 0·15 |
| 12 | 3600 | 159·0 | 210·0 | 90·3 | 0·06 | 0·06 | 0·02 | 0·14 |

Division of elements of column (1) by column (5) and multiplication by the elements of columns (2), (3) and (4) yields the corresponding element of the desire-line tableau. Because there is only one R. and C. destination the white-collar trips are simply equal to their origins.

implicitly from the L.P. land-use solution in Chapter 6 the same job distribution is used. The stratification of the residential locations is simply that the population of each residential zone is divided between industrial and retail—commercial workers in proportion to the total numbers of each class of worker—this case in the ratio of 22,700 to 14,300. With some rounding off the origins and destinations of the work force are as shown below.

The gravity model may now be applied. It is of interest to note that no mathematical difficulties are encountered as the size of the problem is increased. The computational effort increases but the calculations themselves are not increased in complexity.

TABLE 8.3(a). WORK-TRIP DESIRE LINES (INITIAL SOLUTION)

| $o$ \ $d$ | 1 | 6 | 8 | | | $\sum_d T_{od} = R_o$ |
|---|---|---|---|---|---|---|
| | | | Ind. | R. and C. | Total | |
| 2 | 1870 | 260 | 70 | 1500 | 1570 | 3700 |
| 3 | 3020 | 850 | 130 | 2500 | 2630 | 6500 |
| 4 | 930 | 430 | 140 | 1000 | 1140 | 2500 |
| 5 | 1780 | 1680 | 240 | 2800 | 3040 | 6500 |
| 7 | 865 | 310 | 125 | 800 | 925 | 2100 |
| 9 | 410 | 1470 | 120 | 1200 | 1320 | 3200 |
| 10 | 550 | 2750 | 200 | 2400 | 2600 | 5900 |
| 11 | 660 | 270 | 70 | 1400 | 1470 | 2400 |
| 12 | 1500 | 1500 | 500 | 1500 | 1550 | 5000 |
| $\sum_o T_{od}$ | 11,585 | 9520 | 1595 | 14,300 | | |
| $E_d$'s | 9200 | 11,600 | 1900 | | | |

The adjustment factors may now be calculated, viz.

| Destination zone | 1 | 6 | 8 |
|---|---|---|---|
| Adjustment factor | $\frac{92}{116}$ | $\frac{116}{952}$ | $\frac{19}{16}$ |
| Value of $E^1$ | 720 | 1350 | 2250 |

A second iteration yields Table 8.3(b).

TABLE 8.3(b). WORK-TRIP DESIRE LINES FOR ANZAC CITY (FIRST ITERATION)

| o \ d | 1 | 6 | 8 | | | $\sum_d T_{od} = R_o$ |
|---|---|---|---|---|---|---|
| | | | Ind. | R. and C. | Total | |
| 2 | 1810 | 300 | 90 | 1500 | 1590 | 3700 |
| 3 | 2700 | 1150 | 150 | 2500 | 2650 | 6500 |
| 4 | 790 | 550 | 160 | 1000 | 1160 | 2500 |
| 5 | 1400 | 1970 | 330 | 2800 | 3130 | 6500 |
| 7 | 750 | 410 | 140 | 800 | 940 | 2100 |
| 9 | 280 | 1500 | 220 | 1200 | 1420 | 3200 |
| 10 | 420 | 2860 | 220 | 2400 | 2620 | 5900 |
| 11 | 570 | 360 | 70 | 600 | 670 | 1600 |
| 12 | 1250 | 1500 | 750 | 1500 | 2250 | 5000 |
| $\sum_o T_{od}$ | 9970 | 10,600 | 2130 | 14,300 | | |
| $E_d$ | 9200 | 11,600 | 1900 | | | |

Further iterations may be carried out if desired.

TABLE 8.4. DESIRE-LINE PATTERN ARISING FROM L.P. LAND-USE MODEL

| o \ d | 1 | 6 | 8 | $\sum_d T_{od} = R$ |
|---|---|---|---|---|
| 2 | 2750 | | 1250 | 4000 |
| 3 | 3400 | 400 | 2700 | 6500 |
| 4 | 700 | | 600 | 1300 |
| 5 | | 4100 | 2400 | 6500 |
| 7 | 1800 | | 2000 | 3800 |
| 9 | | 2300 | 1200 | 3500 |
| 10 | | 4400 | 2000 | 6400 |
| 11 | 550 | | 150 | 700 |
| 12 | | 400 | 3900 | 4300 |
| $\sum_o T_{od} = E$ | 9200 | 11,600 | 16,200 | |

### *Comparison with Linear Programming Solutions*

The desire-line pattern arising from the L.P. formulation of this problem with stratified origins and destinations but with employment distribution fixed is shown in Table 8.4.

A comparison of the results of the gravity model and L.P. solutions provides the student with much to debate. The following observations may channel his arguments:

(a) The two sets of results are not strictly comparable as the distribution of origins amongst the zones is not the same in both cases. In the gravity model it is predetermined by the separate application of a land-use model, in the L.P. solution it is part of the optimization process. Zone 7 and zone 11 show up the most differently, the others being reasonably comparable. The results of an L.P. desire-line model using a given land-use distribution are presented by McDonald and Blunden(81) in the *Proc. A.R.R.B.* Although the residential trip ends are not the same the difference is not great. The L.P. solution is, however, quite similar to that given in Table 8.4.

(b) The desire-line pattern focusing on zone 8 is comparable in both solutions. This is to be expected, of course, as it is largely determined by the R. and C. trips.

(c) The industrial work-trip pattern is substantially different. In the L.P. solution the work force is very strongly influenced by their proximity to the industrial zone. This is a result often borne out in practice as the work of Carrol and Bell shows.(82,83)

(d) The high degree of discrimination of the L.P. model and its lack of respect for current or past history emphasizes its primary role as a planning tool for investigating long-term alternative plans and strategies.

CHAPTER 9

# Traffic Assignment

## INTRODUCTION

Traffic assignment is the study of how traffic is apportioned between different modes and over a given mode. In the urban context and particularly in respect of passenger assignment it has been common to refer to the apportionment between modes as modal split and over a given mode just simply assignment.

We shall endeavour to use the terms inter- and intramodal assignment to distinguish the two types and we shall consider also the problem of freight as well as passenger assignment.

Traffic assignment presents the engineer, planner or operations manager with a challenging task. In tackling this problem he becomes a genuine transport specialist. It is somewhat paradoxical that the engineer with his long experience at inventing, developing and designing the vast apparatus of transport has paid so little attention to its operational performance in the general socio economic sense.

Traffic operations and management have been based traditionally on clerical and administrative disciplines. This, taken together with the fact that at the turn of the century the rail and sea modes dominated the transport scene, each in splendid isolation, provided a scanty background to the complex interactions of today's transport activity. It is not surprising that the early splutterings of the internal combustion engine provided little indication that 50 years later the cathedral-like grandeur of New York's Grand Central Station would be dwarfed by the giant PAN AM skyscraper or that motorways would spread continuous urban development over thousands of square miles of rural countryside. The insularity of the traditional modes has given way to a vigorous interplay between the old and the new. The stage has thus been set for the operations research spotlight to illuminate this tangled scene.

Two decades ago the early traffic engineers formally recognized the problem, but as their task was primarily that of planning for the use of the private motor car their approach was one of isolating the demand for road transport at the traffic generation stage. This was done by extrapolation of the trends for motor-vehicle use of the relatively short past, by cross-section regression or by assuming mass transport had reached a near-static state and all new demand would be motor-car oriented. There was little or no thought given to the analysis of plans that introduced new mass transport facilities as bona fide inputs. The net effect is that at this point in time intermodal assignment is virtually unexplored.

By contrast, the concentration of effort on road-transport problems has produced a rapid development of techniques in intramodal assignment—from subjective appraisals, through diversion curves to quite sophisticated feedback techniques.

In all this the assignment problem for freight has been left to the powerful forces of business economics and government regulation.

## BASIC CONCEPTS AND CRITERIA

In developing a theoretical foundation for the assignment process, it is important to understand that there is a basic conflict between the transport-system manager and the shipper. The transport-system manager may be a government, a port authority, a railway-traffic manager. The shipper is the person or group wishing to make a trip. The shipper seeks to select a mode and/or route that suits him best, even if his course of action results in disability to other shippers. The system manager, on the other hand, would seek to direct individual actions to ensure the maximum good for all. In this context the management function may be extended to the planning of new facilities. Whilst it is important to recognize that there is a basic conflict, it is the shipper who tends to dominate the situation, at least in a free-enterprise society. This is due to his absolute economic right to choose his method of shipment in a transport context that offers many alternatives.

On the other hand, there is a level at which government regulation or industry agreement have an overriding role in the boundary-condition sense. Rail/road coordination acts, taxation policies, international airline agreements, air fares and routes, to say nothing of speed limits, one-way streets and road traffic regulations.

## THE SHIPPERS' ASSIGNMENT PHILOSOPHY

There is plenty of evidence to validate the following criteria for the selection of modes and routes:

(a) For freight the shipper seeks to minimize his total distribution cost.

(b) The individual traveller seeks to maximize the sum of his tangible and intangible benefits.

The second criterion may be brought into line with the first if it can be assumed that the intangible benefits can be represented by time savings and time made commensurate with the tangibles by giving it a money value.

The distribution cost may be reasonably considered as the envelope of minimum-cost ordinates of a family-cost curve with the value of the freight as the defining parameter.

In the case of intramodal assignment, especially in urban areas, the direct operating costs are approximately proportional to the travel time and as a consequence the assignment criterion can be simplified to one of minimizing travel time.

## THE MANAGEMENT PHILOSOPHY

From the standpoint of general operations and planning policies, management has little option than to follow the shippers' criterion. If a public-transport authority is to win patronage it must offer a "cheaper" trip than the patron can obtain by some alternative means, including his own car. Management may, of course, resort to regulation and control—parking, road-pricing, coordination, for example. However, taken overall it would be fair to say that minimum cost or travel time provides the essential criterion for assignment provided it is remembered that every trip is either a door-to-door or floor-to-floor event. Looked at another way both the shipper and the management/planning agency have criteria which are generally compatible—both seek to minimize their costs. As saturation is approached some divergences appear. However, in the case of urban passenger transport these separate criteria have in recent years come into open conflict on account of the new-found mobility of the individual passenger. The conflict is not only with other modes but with the whole pattern and structure of the urban land-use context. These broader aspects are more properly a subject for the final chapter of this work.

## ASSIGNMENT AND FEEDBACK

Before proceeding to discuss assignment techniques, mention should be made of the feedback implications of the assignment phase in transport and land-use planning.

Up to the assignment stage the land-use pattern, its traffic generation and the desire-line distribution have all been determined for given travel-time inputs. When the traffic so determined is assigned to the transport system new travel times result and these can be used to repeat the planning cycle. Although the conceptual formulation of the process is thus complete, little work has yet been done on the computational formalities and on the stability of the solution process. However, whilst much remains to be found out concerning the behaviour of the mathematical processes involved in the generalized planning process, it is important that the student understand the key role of the assignment process in the determination of travel times and the need to re-assess the previous stages in the light of the newer travel-time values.

## METHODS OF ASSIGNMENT

Although the minimum-time or minimum-cost criterion is explicit enough there is considerable difficulty in applying it to practical problems for, in all but the most simple circumstances, there are complex interactions between and within modes. However, whilst no general theory is applicable to say an entire urban or continental transport system the overall problem may be partitioned into a number of well-defined sub-problems that are amenable to analytical treatment. The methods for dealing with the varied types of problem may be summarized:

(a) Shortest path—time or distance—for a given desire line.
(b) Successively "shortest" paths with capacity constraints.
(c) Equal travel times over competing routes: (b) leads to (c) if the travel times are considered to be flow dependent.
(d) Minimum total travel time or cost in the system or subsystem.
(e) Diversion curves.
(f) Priority rules.

The above methods may all be applied with varying degrees of difficulty to a single desire line. However, when separate desire lines load common links of the transport network complications arise—particularly if the flow dependence of the travel times is to be taken into account. There is no doubt that the minimization of total travel-time approach when formulated as a linear programme provides the most general and powerful specification for the problem. The inability of the L.P. method to take account of variable travel times appears at first restrictive. However, with computer help travel-time effects may be included by adjustment of travel times after each L.P. solution and some iterations of the program.

One complication of the optimizing approach—in road traffic anyway—arises from the fact that the individual is not readily controllable by the optimum plan and hence he tends to press the equal travel-time method into service. More and more, however, there is a need to control the individual road user. The control of time access to motorways, road pricing and arbitrary regulations are examples of current efforts to exercise this kind of control. Even so it must be said that the extreme non-linearity of the travel time or cost-curve forces the individual road user to adjust his own preferences to those of his class as a whole.

There is a considerable literature on assignment[84-91] but a great deal of it is of mathematical interest only. From the point of view of the student the principles of assignment methodology are best introduced by a careful study of the papers of Helly[56] and Tomlin.[85] The one deals with the "physical" models and the latter the linear-programming approach. These works are best illustrated by taking some simple examples.

EXAMPLES

Figure 9.1 indicates a typical situation occurring in urban rush-hour periods. Zones 1 and 3 represent major residential and work zones

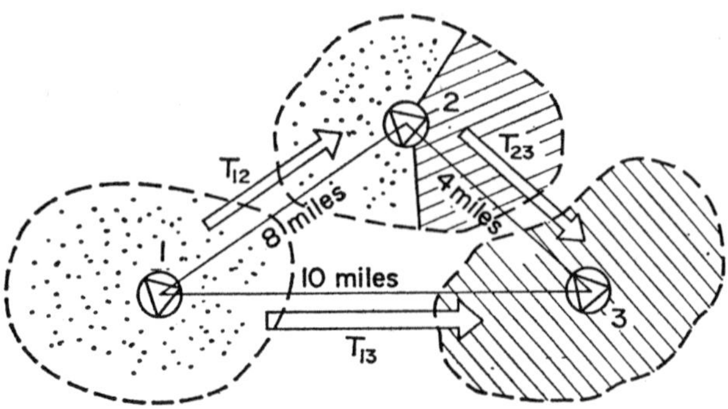


FIG. 9.1. The three-centre assignment problem.

respectively and zone 2 a subsidiary centre with both residential and job potential. The desire-line pattern shown is typical. The road network consists of direct links between the separate centres. The distances between the zone centroids are shown on the diagram (Fig. 9.1). So also are the desire lines. The traffic flow characteristics of the links are as given in Table 9.1.

TABLE 9.1.

| Link | Saturation flow | Zero flow travel time (min/mile) | Level of service factor |
|---|---|---|---|
| 1, 3 | 4000 | 1·6 | 0·5 |
| 1, 2 | 2000 | 2·0 | 1·0 |
| 2, 3 | 2000 | 2·0 | 1·0 |

It will be assumed that $T_{12}$ and $T_{23}$ represent work-trip desires of 1000 vehicle trips/hour. $T_{13}$ will be taken to vary from 2000 to 5000 vehicle trips per hour and the assignment pattern will be studied as $T_{13}$ varies between these limits.

Consider now the various assignment techniques described above. We seek to determine the flows on the links, i.e. $q(1, 3)$, $q(1, 2)$ and $q(2, 3)$.

### *Shortest-route method*

If the shortest route is based on distance then clearly

$$q(1, 2) = 1000 \text{ vehicles/hour,}$$
$$q(2, 3) = 1000 \text{ vehicles/hour,}$$
$$q(1, 3) = 2000 \text{ to } 4000 \text{ vehicles/hour if the capacity constraint is applied.}$$

To accommodate the extra 1000 vehicles/hour there are two possibilities:

(a) To relax the constraint on link 1, 3 and provide extra capacity. In many early analyses this in fact was regarded as the appropriate course to adopt. In other words, the traffic demand was freely assigned with the idea of using the result as the basis for design:

(b) Divert up to 1000 vehicles/hour to links (1, 2), (2, 3).

If the shortest route is to be defined in terms of travel time then a series of computations need to be made.

Table 9.2 gives travel-time values obtained from using the travel-time/flow equation of Chapter 3, viz.

$$t(i,j) = d(i,j)t^0(i,j)\,\frac{1-(1-j)y(i,j)}{1-y(i,j)}\,.$$

Inspection of Table 9.2 shows that a new situation has arisen. If zero-flow travel times were considered then the direct route (1, 3) has a travel time of 16 minutes and the alternative path through 2 takes 24 minutes to complete. In these circumstances the assignment would be similar to that given above for the distance paths. The same situation holds when $T_{13} = 2000$ and $T_{12}, T_{23} = 1000$ (each). However, when $T_{13}$ reaches 3500 vehicle trips/hour the shortest time path from 1 to 3 is via 2, i.e. 48 versus 74·5. It is clear then that when $T_{13}$ approaches 3500 there will be some diversion to the alternative route. Of course the travel time on this route will then increase and some trial and error would be required to find the point of balance. This leads naturally to the equal travel-time method of assignment.

TABLE 9.2.

| Link | $s(i,j)$ | $j$ | $t^o(i,j)$ | $d(i,j)$ | $q(i,j)$ | $y(i,j)$ | $t(i,j)$ |
|---|---|---|---|---|---|---|---|
| 1, 3 | 4000 | 0·5 | 1·6 | 10 | 2000 | 0·50 | 24·0 |
| | | | | | 2500 | 0·63 | 29·4 |
| | | | | | 3000 | 0·75 | 40·0 |
| | | | | | 3500 | 0·88 | 74·5 |
| | | | | | 4000 | 1·00 | |
| 1, 2 | 2000 | 1·0 | 2·0 | 8 | 1000 | 0·50 | 32 |
| | | | | | 1500 | 0·75 | 64 |
| | | | | | 2000 | 1·00 | |
| 2, 3 | 2000 | 1·0 | 2·0 | 3 | 1000 | 0·5 | 16 |
| | | | | | 1500 | 0·75 | 32 |
| | | | | | 2000 | 1·00 | |

### *Equal Travel Times*

The condition for equal travel times may be stated formally

$$t(1, 3) = t(1, 2)+t(2, 3). \tag{i}$$

Now if $f$ is the proportion of $T_{13}$ diverted to the alternative route then the flows in the links are

$$q(1, 3) = (1-f)T_{13}, \tag{ii}$$

$$q(1, 2) = T_{12}+fT_{13}, \tag{iii}$$

$$q(2, 3) = T_{23}+fT_{13}. \tag{iv}$$

Now

$$t(1, 3) = 10\times 1{\cdot}6\,\frac{1-0{\cdot}5q(1, 3)/4000}{1-q(1, 3)/4000}.$$

If we substitute (ii) for $q(1, 3)$ and measure flows in 000's

$$t(1, 3) = 16\,\frac{4-0{\cdot}5(1-f)T_{13}}{4-(1-f)T_{13}}$$

remembering that $T_{12} = T_{23} = 1000$, then

$$t_{12} = 16\,\frac{2}{1-fT_{13}}$$

and

$$t_{23} = 8\,\frac{2}{1-fT_{13}}$$

Substituting these results in (i) gives

$$f^2-f\frac{T-13}{T}-\frac{5T-16}{T^2} = 0.$$

Note: the suffix 13 has been dropped.

This equation may be solved to give values of $f$ for various levels of the traffic demand from 1 to 3, i.e.

| | |
|---|---|
| $T_{13} = 2000$ | $f = 0$ |
| $= 3000$ | $= 0$ |
| $= 4000$ | $= 0{\cdot}10$ |
| $= 5000$ | $= 0{\cdot}20$ |

***Minimum travel time***

The total travel time for the system is

$$V = q(1, 3)t(1, 3)+q(1, 2)t(1, 2)+q(2, 3)t(2, 3)$$

where $q(1, 3)t(1, 3)$, etc., are all functions of $f$ as given above.

The condition for minimum total travel is obtained by putting $dV/df = 0$, i.e.

$$t(1, 3)\,\frac{d}{df}q(1, 3)+q(1, 3)\,\frac{d}{df}t(1, 3) \text{ and similar terms } = 0.$$

This equation yields a cubic in $f$ and a numerical solution gives

$$\begin{aligned} T_{13} &= 2000 & f &= 0 \\ &= 3000 & &= 0{\cdot}03 \\ &= 4000 & &= 0{\cdot}13 \\ &= 5000 & &= 0{\cdot}20. \end{aligned}$$

***Priority to Major Desire Line $T_{13}$***

The amount of travel on route 1, 3 is

$$V = q(1, 3)t(1, 3) + fT_{13}\{t(1, 2) + t(2, 3)\}.$$

Again if we differentiate with respect to $f$ an equation giving the diversion for various values of $T_{13}$ is obtained.

The results are

$$\begin{aligned} T_{13} &= 2000 & f &= 0 \\ &= 3000 & &= 0{\cdot}09 \\ &= 4000 & &= 0{\cdot}16 \\ &= 5000 & &= 0{\cdot}20. \end{aligned}$$

The above examples illustrate a number of important points:

Shortest-time paths with adjusted travel times provide a rational basis for "parallel"-path assignment.

Equal-travel-time assignment is feasible in simple parallel-corridor situations. It is particularly useful for the analysis of by-pass problems.

Optimization of total travel time or travel time for selected desire lines is impracticable in all but the simplest problems.

***Linear Programming***

If the optimization of travel time over a network is to be effectively exploited it is necessary to resort to linear programming methods.

We have already explained, in Chapter 2, the assignment phase of the objective formulation of the overall land-use/transport problem. It will be instructive here to illustrate with an example the L.P. solution of the problem considered above. We will need to make some changes in the definition of the variables and parameters. Figure 9.2 indicates the principal variables and parameters that are required in the analysis. We shall consider the solution for the $T_{13} = 3500$ vehicles/hour with $T_{12}$ and $T_{23} = 1000$ as in the previous exercise. It is necessary to make some assumptions concerning the travel times on the links and it is appro-

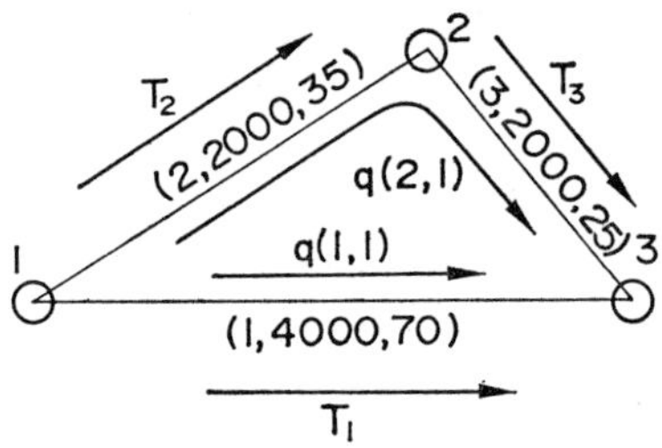


FIG. 9.2. Equivalent network for three-centre problem.

priate to relate them to the flows that are to be expected on the links. In this case, say,

$$t_{13} = 70,$$
$$t_{12} = 35,$$
$$t_{23} = 25.$$

The capacities of the links may be taken as already given. The number sets shown against the links in the diagram (Fig. 9.2) represent in order the link number, its capacity and travel time. From the diagram it is not difficult to construct the land use/transport incidence matrix. There are two possible paths $m = 1, 2$ for each desire line—one by the direct link and the other via the intermediate node (see Table 9.3). The capacity constraint equations may be written down directly by summing the link columns vertically, viz.

$$q(1,1)+q(2,2)+q(2,3) \leqslant 4000,$$
$$q(2,1)+q(1,2)+q(2,3) \leqslant 2000,$$
$$q(2,1)+q(2,2)+q(1,3) \leqslant 2000.$$

The trip-end balance gives

$$q(1, 1)+q(2, 1) = 3500,$$

$$q(1, 2)+q(2, 2) = 1000,$$

$$q(1, 3)+q(2, 3) = 1000.$$

The objective function is obtained by multiplying the flows by the coefficients of the $\sum(t)$ column, i.e.

$$Z = 70q(1, 1)+60q(2, 1)+35q(1, 2)+95q(2, 2)+25q(1, 3)+105q(2, 3).$$

TABLE 9.3.

| Travel times | | | 70 | 35 | 25 | $\Sigma t$ |
|---|---|---|---|---|---|---|
| Flows $q(m, n)$ | Desire line $n$ | $m$ \ $l$ | 1 | 2 | 3 | |
| $q(1, 1)$ | 1 | 1 | 1 | | | 70 |
| $q(2, 1)$ | (3500) | 2 | | 1 | 1 | 60 |
| $q(1, 2)$ | 2 | 1 | | 1 | | 35 |
| $q(2, 2)$ | (1000) | 2 | 1 | | 1 | 95 |
| $q(1, 3)$ | 3 | 1 | | | 1 | 25 |
| $q(2, 3)$ | (1000) | 2 | 1 | 1 | | 105 |
| Saturation flow | | | 4000 | 2000 | 2000 | |

The student may wish to set up the Simplex Tableau with three slack and three artificial variables and solve the problem as a mathematical. The solution is, however, obvious by inspection, viz.

$q(1, 1) = 2500, q(2, 1) = 1000$ and flow on link 1 = 2500,

$q(1, 2) = 1000, q(2, 2) = 0$ and flow on link 2 = 2000,

$q(1, 3) = 1000, q(2, 3) = 0$ and flow on link 3 = 2000.

This is a most unrealistic solution. Travel times should be adjusted and the process repeated. In this problem, however, the solution will oscillate indefinitely if a simple feedback law is used. It is necessary to find a modified feedback function or method that will ensure stability. Such methods are discussed later in the assignment of traffic to the ANZAC City road system. In the meantime we will illustrate the principles involved in ensuring some stability by incremental application of the demand or certain components of it. This is tantamount to putting the network through a peak-hour "experience" until it has adjusted to the total load or sheds some of it.

In the above example we will assume that the minor desires $T_{12}$ and $T_{23}$ are loaded in full initially and $T_{13}$ is raised from 2000 units to 3500 in steps of, say, 500 units. This initial assignment will yield travel times of

$$T_{13} = 24 \text{ minutes}, \quad T_{12} = 32 \text{ minutes} \quad \text{and} \quad T_{23} = 16 \text{ minutes}.$$

It should be clear then that the first increment of $T_{13}$ (i.e. 500 units) will be loaded onto the direct path (i.e. $m = 1$). New travel times will be $t_{13} = 29{\cdot}4$ and $t_{12}$, $t_{23}$ as previously. The next increment will also go to the $m = 1$ path, bringing the load assigned to it to 3000 units. The travel time on (1, 3) is still less than for (1, 2, 3). However, when the next increment is added the situation changes. $t_{13}$ rises to 74·5 minutes. The next increment would go to route (1, 2, 3). But in this particular example the total demand $T_{13}$ is only 3500 units. The assignment process would therefore stop before this last increment was added and a rounding-off error of up to 500 units would need to be accepted. If the $T_{13}$ demand had been 4000 units then the final assignment would have been 3500 units to $m = 1$ and 500 units to $m = 2$ giving a diversion of $12\frac{1}{2}\%$, i.e. $f = 0{\cdot}125$. This agrees well with the theoretical value obtained in the last section for the minimum total travel-time criterion. In the problem as posed $f = 0$ when the theoretical value should be about 0·075. If greater precision for the 3500 maximum demand case is required the load increment should be reduced, say to 100 units after $T_{13}$ reaches 3000 units. It is left to the student to show that this would result in an assignment of 3300 units of $T_{13}$ to route (1, 3) and 200 to route (1, 2, 3). The resulting $f$-value is 0·060 and this also agrees well with the theoretical value of about 0·075.

***Diversion Curves***

The Diversion-curve method of assignment has been widely employed in transport planning exercises over the past decade. Although empirically based it reflects the powerful influence of the equal travel-time

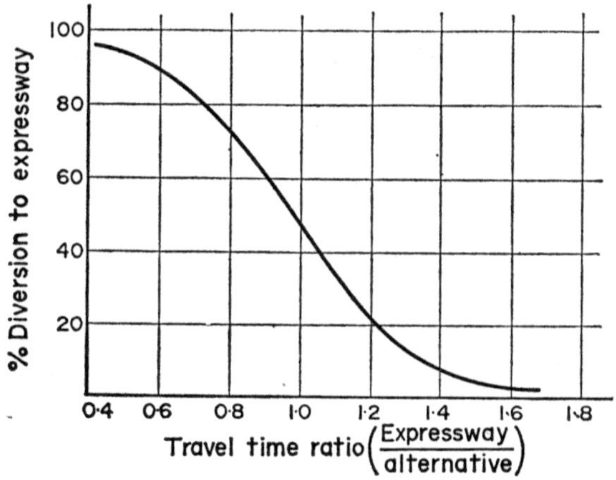


**FIG.** 9.3(a). Bureau of Public Roads-type diversion curve.

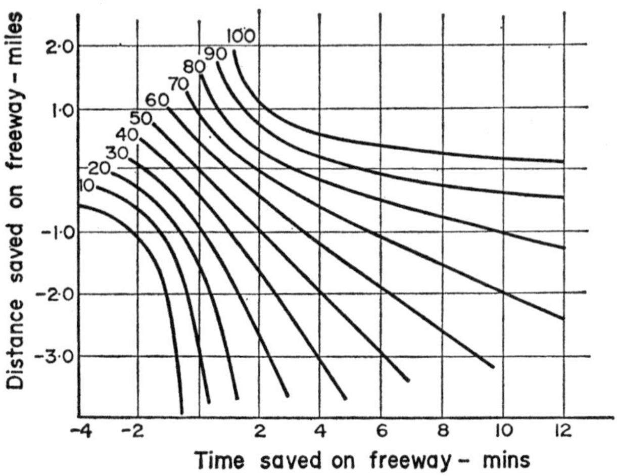


FIG. 9.3(b). Californian Division of Highways diversion curves.

doctrine of the individual driver or shipper. Two well-known versions of the method are due to

(a) the Bureau of Public Roads,
(b) the Californian Highways Division.

The former takes account of relative travel times only, whilst the Californian method bases the diversion on two parameters—absolute time and distance differentials. Figure 9.3 shows the general character of the diversion characteristic of the two methods. Formulae to fit the curves have been developed, viz.

Bureau of Public Roads

$$100f(= \%\ \text{diversion}) = \frac{100}{1+t_R^6}$$

where $t_R$ is the travel-time ratio $\left(\equiv \dfrac{\text{time on new system}}{\text{time on old system}}\right)$.

Californian Method (for diversion to freeways)

$$100f = 50 + \frac{50(d+0{\cdot}5t)}{(d-0{\cdot}5t)^2+4{\cdot}5}$$

where $t$ and $d$ are the time and distance savings on the freeway route measured in minutes and miles respectively.

Note: $f$ is diversion to the new system or the freeway.

## INTERMODAL ASSIGNMENT

This is perhaps the least developed phase of the transport planning methodology. That this is so is due in large measure to the technological and operational insularity, of the main transport modes. Because the motor vehicle is a door-to-door mode the various analyses of road-transport problems have automatically taken this into account. In the other modes direct responsibility for the passenger ceases at the terminal —sometimes with only half of the trip complete! At this point in time it would be true to say that the technical specialists know more of the interaction of their respective modes with land use than with their transport counterparts. This is due, of course, to the non-existence of a technically based professional discipline of transport and also to the fact that in the commercial sense the various modes are in competition and sometimes it is a life-and-death affair.

It has been pointed out earlier in this chapter that minimum cost or minimum travel time are the basic criteria for determining the distribution of traffic over a transport network and this holds whether or not the network is homogeneous or heterogeneous in the modal sense. However, when mixed modes are involved then due account must be taken of modal transfer. It should be stressed again that walking must be considered as a mode and as a result virtually all trips are made on a mixed mode system. In the area of freight traffic the key role of the transfer facilities has been recognized and spectacular progress has taken place in the operational characteristics of freight transfer facilities —containers, bulk handling, roll-on roll-off, piggy back, pipe lines and so on. We shall therefore discuss firstly the problem of intermodal assignment of freight.

### *Freight Assignment*

In Chapter 5 the Distribution Cost equation was discussed and Fig. 5.13 indicates the nature of the relationship between distribution cost and the value of the freight. If now a cumulative distribution of a total freight demand (in ton miles) for a given region or community is constructed and superimposed on the distribution-cost graph it is easy to determine the proportions that should use the various modes. A typical result is shown in Fig. 9.4. Whilst the various curves shown are not based on real data they are nevertheless typical. The frequency function of a community's total freight demand has a roughly exponential characteristic and the cost curves follow the trends shown. The resulting assignment apportions some 50% to sea, 25% each to road and rail and a negligible amount, though economically important proportion, to air. It is interesting to note that the overall distribution of freight between modes for the Australian continent is in similar proportion as Table 9.4 shows. This method is directed, of course, at achieving some kind of optimum distribution for the region at large. The assignment problem for the individual shipper is one of minimizing his own distribution cost. Provided that the system is not working too close to saturation this becomes essentially a matter of setting up minimum-cost paths. In practice, however, congestion at ports, peak overloads, general road-traffic difficulties can cause situations just as demanding of diversions as is the case in ordinary car-traffic assignment. It is unfortunate that up until now there has been little serious analytical study of freight-traffic congestion.

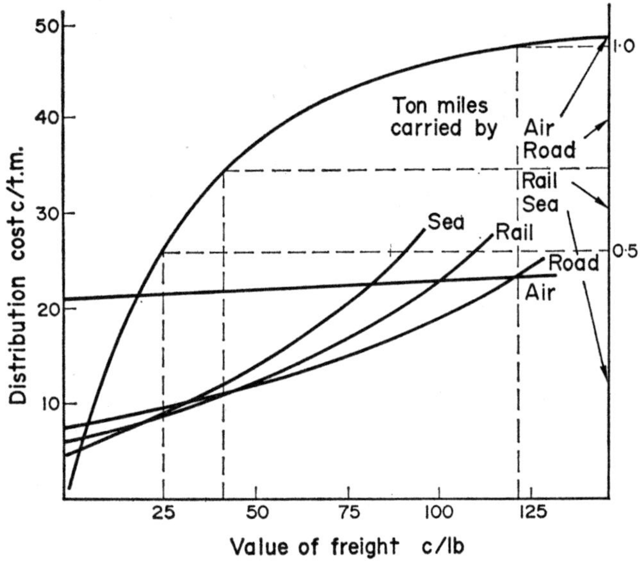


FIG. 9.4. Assignment of freight between main modes.

TABLE 9.4. ESTIMATED DOMESTIC FREIGHT TASK PERFORMED BY EACH MODE—AUSTRALIA 1960[49]

| Mode | Tons carried, millions | % of total tons | Ton miles millions | % of total ton miles |
|---|---|---|---|---|
| Sea | 19 | 4·8 | 18,855 | 49·1 |
| Road | 306 | 77·2 | 11,125 | 29·0 |
| Rail | 71 | 19·2 | 8,357 | 21·8 |
| Air | 0·06 | — | 29 | 0·1 |

***Intermodal Passenger Assignment***

The intermodal assignment problem in urban areas has become a major topic for debate but at this point in time little real progress has been made in developing workable solutions to it.

Broadly two approaches have been taken:

(a) determination of the modal distribution before desire-line pattern is obtained,
(b) after desire-line determination.

The former is the one that has been more or less universally employed up to the present. It has merit—particularly as there is an obvious strong correlation between modal choice and vehicle-ownership levels in the origin land-use zones. It is also of relevance to observe that in the absence of constraints the cost of private-vehicle travel is highly competitive with mass modes. This is particularly true if ownership costs are ignored as well they may be in affluent societies. And if ownership costs are included and if the load factor reaches 3 or 4 private car travel is still competitively cheap.

Even so the approach is dangerous—both analytically and practically. From the analytical point of view it is only after the desire line has been determined that it is possible to assess the comparative merit of alternative modes. Mass transport demands high-intensity desire lines in order to permit a frequency of service that will cut waiting time to acceptable levels. Further, the *od* link needs to be specified so that the constraints imposed at the destination end can be properly related to the trip origins. The destination end constraints are extremely powerful—road capacities as centres of concentration are approached and parking limitations. There is little need to look for proof beyond noting that the proportions of workers and others who come to the central areas of cities like London, Sydney, New York, Tokyo are of the order of 80–90%.

With due account then to the constraints, what analytical procedure should be adopted for intermodal assignment?

It would seem that a travel-time or cost criterion is just as meaningful in this situation as it is in intramodal assignment, provided that door-to-door costs or travel times are used. If the cost variable is to be used then a suitable money value must be given to the travel time.

An appropriate formula for calculating the door-to-door travel time, $t$, for a mass transit journey is

$$t = \frac{L_f \,.\, C}{2T} + \frac{d \,.\, R_t}{v_t} + W$$

where $L_f$ is the load factor of the vehicle,
$C$ is its capacity (persons),
$T$ is the intensity of trip demand—persons/unit time,
$d$ is the direct distance between transit terminals,
$v_t$ is the average operating speed (including stops),
$R_t$ is the transit service-route factor,
$W$ is the average time spent at both origin and destination ends in concentrating on the terminal, i.e. walking time or time spent in other modes.

An example will serve best to further discussion on this difficult problem.

EXAMPLE

An important transport corridor connects a major residential zone to the central area of a large metropolis. The number of daily commuters to the centre in the rush hour is 10,000. Of these, 30% are junior office workers (mainly female) and 10% are executives and professional specialists who have special need of a car in their work. The direct distance between the zone centroids is just 10 miles. The journey time by car, $t_c$, may be described by some adaptation of the travel-time/flow relationship 3·22, viz.

$$t_c = t_c^o \frac{1-(1-j)q/S_e}{1-q/S_e} + w$$

where $t_c^o$ is the overall travel time from home to the city parking point under free flow (light traffic) conditions; it is a quantity readily obtained by direct measurement and it takes account of intersections, speed limits and other statutory controls, nature of the environment and so on; it is true also that $t_c^o$ is not difficult to estimate—

$$t_o^c = \frac{DR}{V} + nc(1-\alpha)^2/2,$$

here

$D$ is the direct distance between a given origin and destination,

$R$ is the route factor for the journey,

$V$ is the average running speed,

$n$ the number of major intersections,

$c$ a representative cycle time,

$\alpha$ the proportion of green time favouring main-road traffic under conditions of light traffic.

Assume in this example

$D$ = 10 miles,

$R$ = 1·12,

$V$ = 40 m.p.h.,

$n$ = 10,

$c$ = 1/60 hours,

$\alpha$ = 0·75,

then $t_c^o = 0{\cdot}28 + 0{\cdot}05 = 0{\cdot}33$ hours, i.e. 20 minutes—a very reasonable estimate of the time to travel 10 miles on the major-road system of an urban area with negligible other traffic.

$j$ is the level of service factor; for rush-hour car traffic on the arterial road system a suitable value would be 0·50.

$S_e$ is the equivalent saturation flow for the corridor; an estimate of this parameter requires certain simplifying assumptions based on professional expertise but such an estimate can be usefully guided by the application of the maximum flow/minimum cut theorem of network theory; here we will assume that $S_e$ is 2500 vehicles/hour.

$w$ is the average time taken in moving between parking point and work place; here we will assume $w = 6$ minutes.

Now the non-captive demand (for both public and private transport) is 6000 person trips in the rush hour. Let $T_s$ and $T_c$ be the numbers that will use public transport (assumed to be a rapid transit system) and car respectively. The travel time for trips by car is

$$t_c = 0{\cdot}33\,\frac{1-(1-0{\cdot}5)q/2500}{1-q/2500}+0{\cdot}10,$$

but

$$q = (T_c+1000)/p$$

where $p$ = vehicle occupancy factor; assume $p = 1{\cdot}2$ and the second term of the bracket is the number of captive trips for car commuters; thus

$$t_c = 0{\cdot}33\,\frac{1-0{\cdot}5(T_c+1000)/3000}{1-(T+1000)/3000}+0{\cdot}10$$

$$= \tfrac{1}{6}\,\frac{S-T_c}{2-T_c}+0{\cdot}10$$

if $T_c$ is measured in 000's of trips/hour.

For the public-transport patrons

$$t_t = \frac{L_f\,.\,C}{2\,.\,T_t}+\frac{D\,.\,R_t}{V_t}+W.$$

Here $L_f = 1{\cdot}25$,

$C = 800$ persons,

$R_t = 1{\cdot}10$,

$V_t = 35$ m.p.h.,

$W = 18$ minutes.

Therefore

$$t_t = \frac{1\cdot25\times800}{2(T_t+3000)}+\frac{10\times1\cdot1}{37}+0\cdot30$$

$$= \frac{1}{2(T_t+3)}+0\cdot60,$$

again $T_t$ is measured in 000's and account is taken of the 3000 captives to public transport.

Equating travel time by public and private transport yields the following

$$\frac{1}{2(T_t+3)}+\tfrac{1}{2} = \tfrac{1}{6}\cdot\frac{5-T_c}{2-T_c}$$

but

$$T_t+T_c = 6(000\text{'s})$$

and we may thus write an equation for $T_c$

$$\frac{1}{g-T_c}+1 = \tfrac{1}{3}\cdot\frac{5-T_c}{2-T_c}$$

the valid solution of which is

$$T_c = 750.$$

This results in a very encouraging state of affairs so far as public transport is concerned for the modal split components would be 8250 and 1750 for transit and cars respectively.

Now consider the effects of improving the road facilities—say to motorway standard. Leaving public-transport and terminal parameters as before and introducing flow parameters appropriate to the motorway, then

$$t_c^0 = 0\cdot25 \text{ hours},$$

$$j = 0\cdot20,$$

$$S_e = 4000 \text{ vehicles/hour}$$

(these values take account of the fact that not all of the car commuters' journey will be on the motorway—80% perhaps).

It is left to the student to introduce these new values into the above equations and solve them. He will find, however, that the situation has changed dramatically for the new value of $T_c$ is 2960 and the split is now 6000 to 4000. By contrast, if elaborate improvements were made to the public transport such as, say, doubling the frequency of service, the effect in the first situation (i.e. the non-motorway case) would result in only 140 new passengers.

The above analysis permits one to make one or two important generalizations. First it would be fair to say that unless constrained by flow parameters there are few situations where the non-captive part of the demand would do other than use cars. An even more powerful constraint is imposed by parking conditions at the destination end. If in the second case considered there were, say, 2000 car spaces available to the commuters in this particular corridor then the model split would be 7600/2400. Those that do travel by car would, of course, have a more pleasant journey. Second, the results of the above calculations show that in the unequal "battle" for the commuters that as demand increases the performance of public transport improves whereas that of individual transport worsens.

In these examples we have used travel time as the dependent variable. A cost function would seem more logical. However, this raises the awkward problem of putting a money value on time. On the other hand, it has been established for urban conditions that transport operating costs are approximately proportional to time and there is thus some real virtue in working throughout with time.

## TRAFFIC ASSIGNMENT FOR ANZAC CITY

The application of the above principles may be illustrated by carrying out an assignment of the work-trip desire lines already calculated for ANZAC City. Reference will also be made to some of the more important other methods of road-traffic assignment in current use.

In this exercise we will assume that the intermodal assignment has been carried out separately. The following distribution of the remaining vehicular trips is obtained from Table 8.2 by arbitrary adjustment of vehicle occupancy figures for individual desire lines.

TABLE 9.5. CAR TRIPS BY ANZAC CITY WORK FORCE

| o \ d | 1 | 6 | 8 |
|---|---|---|---|
| 2 | 600 | 100 | 480 |
| 3 | 870 | 430 | 900 |
| 4 | 400 | 340 | 580 |
| 5 | 350 | 890 | 1000 |
| 7 | 420 | 230 | 530 |
| 9 | 50 | 550 | 490 |
| 10 | 60 | 990 | 730 |
| 11 | 270 | 170 | 350 |
| 12 | 200 | 360 | 610 |

It will be further assumed that the travel times will be determined by the transport impedance function 3.22. The parameters for the road system hierarchy of ANZAC City are given in Table 9.6.

TABLE 9.6.

| Road type | $s$ veh./hour | $t^0$ min/mile | $j$ |
|---|---|---|---|
| Motorway | 4000 | 1·0 | 0·05 |
| Arterials | 3000 | 2·0 | 0·50 |
| Feeder and local streets | 2000 | 3·0 | 1·00 |

A careful investigation of the application of a variety of assignment techniques to ANZAC City was carried out by Farrington.[92]

The first step is to construct the land-use/transport-incidence matrix $A(l, m, n)$. To do this the links of the road network must be numbered. Further it is desirable to simplify the network if possible. In urban networks it is desirable to arbitrarily eliminate local streets to ensure that the main interzonal movements can be accommodated on motorways, arterials and feeder roads. In rectangular grid systems it is difficult to prevent through traffic spreading as there are many equal-length

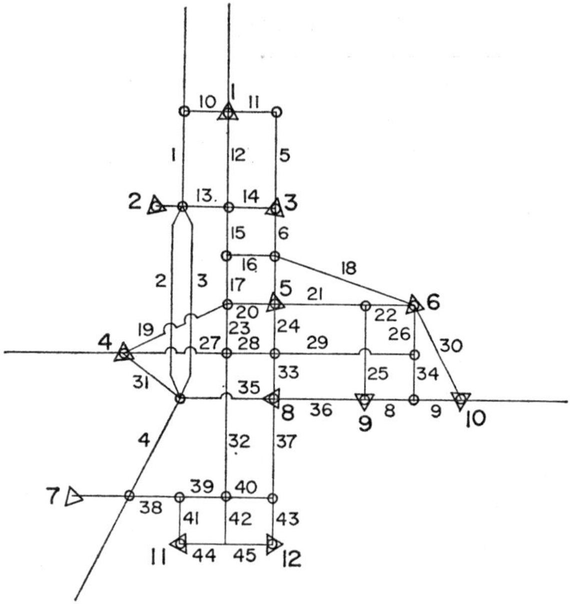


FIG. 9.5. Simplified road network for ANZAC City.

TABLE 9.7. PART OF $A(l, m, n)$ MATRIX FOR ANZAC CITY

| Travel time | | | 1 | 2 | | 1.1 | | 1.5 | | 3 | 1.5 | | 1.5 | | 1.5 | | 1.5 | | 4.5 | | 1.5 | 1.5 | $\Sigma t_l$ |
|---|---|---|---|---|---|---|---|---|---|---|---|---|---|---|---|---|---|---|---|---|---|---|---|
| q | n | m \ l | 1 | 2 | | 4 | | 10 | | 12 | 13 | | 15 | | 17 | | 23 | | 32 | | 38 | 39 | |
| q(1,1) | 1 | 1 | 1 | | | | | 1 | | | | | | | | | | | | | | | 2.5 |
| q(2,1) | (2,1) | 2 | | | | | | | | 1 | 1 | | | | | | | | | | | | 4.5 |
| ... | ... | ... | . | . | | . . | | . | | . . | . . | | . . | | . . | | . . | | . | | . . | . . | ... |
| ... | ... | ... | . | . | | . . | | . | | . . | . . | | . . | | . . | | . . | | . | | . . | . . | ... |
| q(1,13) | 13 | 1 | 1 | 1 | | 1 | | 1 | | | | | | | | | | | | | | | 5.6 |
| q(2,13) | (7,1) | 2 | | | | | | | | 1 | | | 1 | | 1 | | 1 | | 1 | | 1 | 1 | 13.5 |
| ... | ... | ... | . | . | | . . | | . | | . . | . . | | . . | | . . | | . . | | . | | . . | . . | .. |
| Saturation flow | | | 4 | 4 | | 4 | | 2 | | 2 | 2 | | 2 | | 2 | | 2 | | 2 | | 2 | 2 | |

paths between "diagonal" $o$ and $d$ pairs. However, turn penalties come to the rescue in some circumstances. A simplified network for ANZAC City is shown in Fig. 9.5. It has forty-five links.

The next step is to define the shortest routes and as appropriate a number of alternative paths. This is quite a formidable undertaking even for a small network such as the one under discussion. For the purpose of illustration a section of the $A(l, m, n)$ matrix is shown in Table 9.7.

We can now illustrate a number of assignment techniques.

### *Diversion Curves*

Consider desire line $T_{21}$, i.e. $n = 1$.

Travel time by motorway (i.e. $m = 1$) = 2·5.

Travel time by surface streets ($m = 2$) = 4·5.

Therefore $T = 2{\cdot}5/4{\cdot}5 = 0{\cdot}56$,

$T^6 = 0{\cdot}04$

and % diversion $= 96{\cdot}5$

or $T_{21} = 600$

and $q(1, 1) = 580$ $\quad q(2, 1) = 20.$

For $T_{71} = 420$ $\quad T = 0{\cdot}4$ and $T^6 \to 0$.

Therefore all trips will use motorway, i.e.

$$q(1, 7) = 420 \quad \text{and} \quad q(2, 7) = 0.$$

### *Shortest-path Assignment*

The $A(l, m, n)$ matrix shows that $q(1, 1)$ and $q(7, 1)$ are the flows on the shortest path. Therefore

$$q(1, 1) = 600$$

and

$$q(7, 1) = 420.$$

Although the shortest-path method appears naïve, it provides the basis for a number of useful iterative methods of obtaining practical solutions to large-scale assignment problems. The Chicago, Pittsburgh, Bureau of Public Roads and Traffic Research Corporation methods are essentially similar and are based on the following routine.

1. Shortest-time paths are constructed using zero or relatively free-flow travel time (say $y = 0{\cdot}7$).
2. A shortest-path assignment is then carried out and new travel times calculated.
3. Various methods are used to proceed to the subsequent iterations:
   (a) the new travel times are used to redefine the shortest path;
   (b) the weighted average (by flow) is used as the travel time for the next iteration;
   (c) the traffic is distributed between the old and the new paths in inverse proportion to the travel times;
   (d) the old and new volumes are averaged for use in the travel-time/flow equation;
   (e) the McLaughlin method which uses a network algorithm to effect successive iterations.

### *Overall Capacity Constraint*

If capacity constraints are to be imposed either discretely or by virtue of a rational travel-time/flow relationship it is necessary to check the network to ensure that there is sufficient "overall" capacity to permit a feasible solution. If certain desire lines exceed the capacity of the set of $m$ possible paths, that particular desire line must be reduced as in fact it is in practice by virtue of the arguments set out in Chapter 2. Further, the link or links that become saturated must be eliminated from any possible paths for other desire lines. An algorithm for doing this can be constructed from the land-use/transport incidence matrix. Alternatively, if all origins are joined to a master origin and similarly for destinations the maximum flow–minimum cut algorithm may be applied. If maximum flow is greater than $\sum T_{od}$ then a valid assignment is possible. This letter requires simultaneous desire-line and assignment determination.

### *Results for ANZAC City*

Table 9.8 summarizes the results obtained by Farrington in applying various of the above methods to ANZAC City. The table includes only links in the motorway and the two major arterials.

TABLE 9.8. MOTORWAY AND MAJOR ARTERIAL LINK FLOWS—ANZAC CITY

| Link no. | Capacity check | Diversion curves | Direct path | Chicago | McLaughlin |
|---|---|---|---|---|---|
| 1 | 2000 | 1770 | 2000 | 1690 | 1243 |
| 2 | 1400 | 1195 | 1400 | 1090 | 864 |
| 3 | 480 | 440 | 480 | 480 | 777 |
| 4 | 1720 | 1810 | 2170 | 1220 | 1153 |
| 5 | 1220 | 1345 | 1220 | 980 | 866 |
| 6 | 1780 | 1945 | 1780 | 1440 | 1355 |
| 7 | 1780 | 1945 | 1780 | 1010 | 1031 |
| 8 | 1570 | 2100 | 2100 | 1640 | 1463 |
| 9 | 1670 | 1780 | 1780 | 730 | 1161 |
| 24 | 1900 | 2065 | 1900 | 1900 | 1603 |
| 33 | 1900 | 2065 | 1900 | 1900 | 1873 |
| 35 | 2000 | 2100 | 2450 | 1590 | 1700 |
| 36 | 1920 | 2090 | 2090 | 1580 | 1781 |
| 37 | 1250 | 1315 | 970 | 1320 | 1438 |
| 43 | 1450 | 1435 | 1170 | 1320 | 1274 |

In making general comment on his results Farrington considers that the McLaughlin and Chicago methods have merit on the score that they distribute the demand more evenly over the whole network. In Table 9.8 this shows up as a consequence of the lower flows on the links of the major road system. His interpretation is, however, somewhat erroneous. One of the principal criteria of good road planning in the overall land-use/transport context is to ensure that a minimum of interzonal traffic finds its way on to local streets. But the more dispersed the assignment becomes the greater the total amount of travel (passenger or vehicle miles).

## CHAPTER 10

# Evaluation Review and Implementation

### INTRODUCTION

The final phase of the land-use/transport system analysis is the evaluation of the solutions that result from physical planning process. Some form of evaluation procedure is essential for the simple reason that in large-scale problems of this kind there is no unique physical solution. One can borrow from the mathematical programming vocabulary—there are many possible solutions and a lesser number of feasible solutions and of the various feasible solutions the planners' task is to select the best. The object of the physical planning phase is to distinguish the feasible solutions and that of the evaluation phase to select the best one.

The determination of the best or optimum solution immediately calls for some criterion by which to effect the ranking and selection procedure. It also requires some definition of a time horizon on which to arrange the exhibits (i.e. the solutions) for comparison.

We shall consider both these aspects in some detail.

### THE CRITERION PROBLEM

Even though the problems of land use and transport directly or indirectly influence a wide range of community affairs, there are other major areas of community activity that might well be considered as largely independent in the functional sense anyway, e.g. education, health, social services, power, telecommunications, water conservation and so on. In the private sector of the economy rural production, basic industry, housing generally are intimately related to the land use/

transport context. Further tertiary industry is much influenced by the transport system.

The first task facing those concerned with evaluation is to partition the problem and develop consistent lower-level criteria that have direct and meaningful application to the land-use/transport system itself. Just how much of the total resources of the community should be allocated to land-use and transport works is a matter of high-level government policy. However, the relative independence between the major areas of public spending does permit some formal top-level capital budgeting to be undertaken. In the context of our specialized interests we must seek a system with a general disposition of major land uses such as to minimize the amount of essential travel and a transport system which is efficient in itself. Along with these overall objectives are many contingent lower-level criteria—local land-use zones should be protected from major through-traffic movements, the transport system should be safe, pedestrian amenity should be high, parking facilities and transfer terminals adequate, public transport fully exploited, smog and noise properly controlled, high standards of civic design and appropriate open space, viable central city area and lively regional centres.

It is, of course, well nigh impossible to apply multi-criterion approach to other than the most simple problems unless the criteria are compatible and consistent. However, a little thought will show that many of the sub-criteria mentioned above are in fact compatible with a good physical solution of the land-use/transport problem. High-density land uses make public transport more competitive with the motor car, motorways preserve the integrity of local land-use areas and give greater scope for civic design, efficient transfer terminals improve public-transport usage for commuters to central city areas. Even pollution could be much alleviated by the development of an effective electric car for urban use.

It can be seen therefore that a first step in the optimization procedure can be achieved by regarding the lower-level "criteria" as boundary conditions (or goals) in the evaluation phase. The final optimization can then be effected by a maximizing or minimizing process depending on whether the main criterion is a desirable or undesirable measure—we would maximize benefits, legitimate desires or minimize costs, vehicle miles, accidents. In the mathematical programming approach the dual may be used to test the sensitivity of the constraint criteria.

Because planning is only truly meaningful if implementation is to follow and because implementation requires investment, the final evaluation is essentially an economic process.

## TIME HORIZONS

The time-horizon concept in planning and evaluation needs to be fully understood. There has in the past been a great deal of confusion in this matter.

### *The Economic Evaluation Horizon*

In an economic type of evaluation a period of analysis of, say, 10–15–20 years is used as an interval through which to calculate the benefit and cost streams and it is a generally accepted practice to discount them to the present or some time point, usually near to the present, at which the project under construction is expected to come into operation. Conceptually it arises from the commercial norm of recovering the invested capital over a period of a decade or two.

### *The Planning Period*

This is quite a different concept and a much more nebulous one. This planning period seeks to define a future point around which a master plan takes final shape. Not only does it allow the time for longer-term goals to be realized but it provides for some kind of moratorium over the activity that would take place during the period itself—in particular highway authorities define future road reservations and civic authorities frame regulations and policies to guide long-term development towards the final goal, e.g. floor-space ratios, green-belt requirements, new civic features, and so on.

The planning associated with this latter kind of time horizon has traditionally embraced the projective planning methodology. Trend projections and regression models are used to extrapolate the variables to the end of the planning period. The approach is essentially open-ended and there is little concern with implementation as such. On the other hand, the economic horizon is in essence geared to implementation —implementation is in fact the acid test of the plans themselves. If they cannot be put into effect with the budget available, then they may as well be dismissed and considered afresh later in the light of the actual changes that have taken place. In this approach we are in fact always planning in current time even though the future consequences of the current development are discounted and used to establish its "merit" *vis-à-vis* other alternatives.

There are advantages and disadvantages in each of the approaches and one way to obtain something of the best of two worlds is to set up the essential goals as boundary conditions for the various analytical processes.

We might summarize the above brief review of the evaluation problem by saying that the evaluation model of the overall planning process will be developed as an economic model and will be based on the now well-established principles of economic analysis.

## ECONOMIC ANALYSIS OF TRANSPORT PLANS

Most of today's transport schemes are public-investment projects. Urban motorway systems, state highways, new mass transport systems, airports and docks are financed by governments and semi-government bodies. The heavy cost of major transport undertakings and the pressing demands for government funds for the many other requirements of the highly developed modern state have resulted in widespread application of social-welfare economics in government budget planning. The pioneering work of the U.S. Bureau of Reclamation[93] on the economic evaluation of water-resource projects has provided a foundation for the development of economic evaluation methods for many areas of public investment including transport. The A.A.S.H.O. *Manual on Benefit Cost Analyses for Highway Improvements*[94] marked the beginning of formal evaluation procedures in the area of road planning. This manual originated from the pioneering work of the Oregon State Highways Department. Other important contributions have been made by the Californian Highways Division, The Road Research Laboratory in

Great Britain[95,96] to the practical application of formal economic analysis. Recent work on this subject by the new Australian Commonwealth Bureau of Roads is notable.[97,98]

It should be noted, of course, that some kind of economic and financial analyses have been carried out as a matter of course in the various commercially oriented transport undertakings—railways, air transport, shipping. In these cases the analysis is relatively straightforward as the benefit stream can be defined in terms of real cash flows resulting from charges for service. In the more general field of transport activity the definition of direct and indirect benefits is an extremely difficult task. The estimation of investment costs is in general not too difficult.

In Fig. 1.9 (Chapter 1) it will be seen that the evaluation model is located in the feedback channel between the output of the physical planning process and the input of the variable of analysis, i.e. a proposed transport plan. In some situations the evaluation model may be framed in terms of physical measures of performance, i.e. speeds, capacities. However, such measures have a considerable degree of arbitrariness and as a result it is generally true to say that an economic evaluation model is more versatile and meaningful.

## METHODS OF ECONOMIC EVALUATION

There are three basic methods of making an economic evaluation:

(a) Minimization of costs in achieving certain goals and standards.

(b) Maximization of the effectiveness of a system for a given expenditure.

(c) Optimization of some function of benefits *and* costs associated with a project or project set.

The two former methods lead essentially to cost effectiveness measures and are of value when it is difficult or impossible to quantify the benefits in units commensurate with the costs—for example defence, education, social-services programmes. Even in road safety where it is common to put a money value on accidents there may be merit in comparing, say, the effectiveness of seat belts or, say, median dividers in highways on the basis of cost/accident or cost/injury accident.

The third method is the more commonly espoused in transport programmes. Benefits usually take the form of transport-user cost-savings and whilst this definition of benefits is relatively clear cut when applied to rural road improvements its adaptation to the urban scene leaves out of account many vital consequences of road improvements such as effects on public transport, urban sprawl, pollution and so on. It may be better to simply accept a "plan" and seek out the most economical way of implementing it. But having made this observation we will proceed to discuss the formal methodology of benefit/cost evaluations.

### *The Benefit/Cost Function*

There are a number of benefit/cost indices or functions that have usefulness in economic evaluation, viz.

(a) Benefit/cost ratios.
(b) (Benefits—costs).
(c) Transportation cost.

It should be noted that benefits arise by differencing the user or operational costs of alternative schemes or in the case of a single improvement taking the difference between the costs on the new project or the costs on the one it is to replace or supplement.

The benefit ratio takes a number of forms:

$$\frac{\text{Present value of benefits}}{\text{Investment cost}},$$

$$\frac{\text{Equivalent annual benefits}}{\text{Annual capital cost}},$$

$$\frac{\text{First year benefits}}{\text{Investment cost}},$$

$$\frac{\text{Median year benefits}}{\text{Annual capital cost}}.$$

With the exception of the two latter ratios which are special indices devised by the Road Research Laboratory and A.A.S.H.O. respectively, the other benefit ratios and functions give rise to the rate of return. The rate-of-return concept is a commercial adaptation and whilst it is basic to the methodology of formal economic analysis it does present problems in the public-expenditure field because it is in essence connected with the commercial norm of recovering the capital over a finite number of years. Irrespective of whether it is sound or not for a public body to recover its capital outlays in a strictly financial sense, the awkward consequence that arises is the need to predict the traffic characteristics into the 20-year future just to set up a stream of somewhat "dubious" benefits.

It should be pointed out that benefit/cost functions do in general yield a rate of return or discount rate and these indices are of importance in capital budgeting.

### *Rate of Return and Discounting*

Two important rates of return need to be defined:

the Internal Rate of Return,
the Marginal Rate of Return.

The first may be calculated by establishing the discount rate at which the benefit and cost stream are equal. The second is the internal rate of return at which the budget is fully utilized when the projects are ranked in order of their internal rates of return.

If the benefits from a project in a succession of years 1, 2, ... $n$ are $B_1$, $B_2$, ... $B_n$, then the present value of the benefits is

$$PVB = \frac{B_1}{(1+d)} + \frac{B_2}{(1+d)^2} + \dots \frac{B_n}{(1+d)^n}.$$

If this stream of benefits is to accrue from a single investment $I$ then the net present value or present worth of the project may be defined

$$PW = PVB - I.$$

If the annual benefits are constant year by year then

$$PVB = B\frac{1}{1+d}\cdot\frac{1-\dfrac{1}{(1+d)^n}}{1-\dfrac{1}{(1+d)}}$$

$$= B\frac{(1+d)^n-1}{d(1+d)^n} = B/F(n, d)$$

where $F(n, d)$ is the capital recovery factor. This is a well-documented function.

If the benefits increase annually by a constant amount $A$, then the present value of the stream is

$$\frac{B}{F}+\frac{A}{F}\left\{\frac{1+d-nf}{d}\right\}$$

where $f$ is the sinking fund factor $= F-d$.

## Examples

Determine the internal rate of return for the following projects, each costing \$1,000,000:

(a) annual benefits of \$80,000 for 25 years,

(b) \$60,000 initial benefits and then increasing by \$5000 per year for 20 years.

For (a)

For the discount rate to equal the internal rate of return

$$PVB = I = 10^6,$$

$$\therefore\ F(25, d) = B/I = 0{\cdot}08.$$

Inverse interpolation in a table for $F$ for $n = 25$ gives

$$d = 6{\cdot}3\%.$$

For (b)

$$PVB = \frac{60\times 10^3}{F(20, d)} + \frac{5\times 10^3}{F(20, d)}\frac{1+d-20f(20,d)}{d} - 10^6.$$

We must proceed by trial and error:

when $d = 5\%$ $F(20, 5) = 0{\cdot}08$ and $f(20, 5) = 0{\cdot}03$

$$PVB = 1{\cdot}3\times 10^6,$$

when $d = 8\%$ $F(20, 8) = 0{\cdot}102$ and $f(20, 8) = 0{\cdot}022$

$$PVB = 0{\cdot}98\times 10^6.$$

Therefore the internal rate of return is just less than 8%.

If now there were a number of other projects with internal rates of return of 6% or less and an available budget of $\$2\times 10^6$ the marginal rate of return can be determined (Table 10.1).

TABLE 10.1.

| Projects | Investment cost $I | ΣI | I.R.R. % |
|---|---|---|---|
| B | $10^6$ | $10^6$ | 8 |
| A | $10^6$ | $2\times 10^6$ | 6·3 |
| ------ | ---- Budget | line ---- | ------ |
| C } | Other projects | | 6 |
| D } | | | 5 |

The marginal rate of return is 6%.

If now the projects A and B are discounted at 6% their present worths will be:

for A

$$PW = \frac{80\times 10^3}{F(25, 6)} - 10^6 = 0{\cdot}03\times 10^6$$

$$= \$30{,}000,$$

and for B

$$PW = \frac{60\times 10^3}{F(20,\ 6)} + \frac{5\times 10^3}{F(20,\ 6)}\,\frac{1+0{\cdot}06-20f(20,6)}{0{\cdot}06} - 10^6$$

$$= \$190{,}000.$$

DETERMINATION OF BENEFITS AND COSTS

In the land-use/transport context the definition and quantification of benefits is an extremely difficult task. Although benefits are often defined as the difference between two cost terms, i.e. cost of operating on a new road compared with that for an old facility, the term costs in benefit/cost jargon usually refers to the investment and maintenance costs relating to the permanent or fixed facilities. In such cases little difficulty is experienced in specifying them in monetary units.

To return to benefits or disbenefits as such we can define

(a) direct benefits to the user, i.e. reduced costs, savings in time, reduction of accidents;

(b) induced benefits due to extra total traffic, reduction of traffic on local streets;

(c) disbenefits, i.e. pollution, over crowding, urban sprawl, loss of amenities.

At this point in time little success has been achieved at quantifying other than the direct benefits to the transport user and the shipper. Even in this regard there has been little exploitation of the direct benefits except in the case of users of motor cars on the intercity highway.

In this situation the problem is quite clear cut. The method of approach is straightforward. The annual road-user cost $C$ for rural roads may be expressed as

$$C_R = 365\times ADT\times \sum (d_g\times c(v,\ g))$$

where $ADT$ is the average daily traffic,

$d_g$ is the lengths of road in various grade classes,

$c(v,\ g)$ is the unit operating cost of a vehicle as a function of speed and grade.

$v$ is generally taken as a representative operating speed for the whole year's traffic operations and defined as a function of a representative flow (i.e. 30th highest hour) to the saturation flow.

More recent work has attempted to integrate the product of the hourly flow and the cost function over the whole year, i.e.

$$\sum_{g} \int_{0}^{8760} q(t)c_g(q)dt.$$

$c(q)$ is obtained by substituting $q$ for $v$ from some appropriate speed/flow relationship.

In the urban scene the situation is much more complex not only because of the difficulty of defining a meaningful operating speed but also because the level of flow-causes changes in its distribution over the road system. The road-user costs resulting from an urban road improvement may be written

$$C_U = \sum_{h=1}^{8760} \sum_{l=1}^{L} q_l c_l.$$

In this case $c_l$ is better considered as a function of the travel time on the link—see Fig. 5.10 of Chapter 5.

Both $C_R$ and $C_U$ need to be evaluated for the new facility and the old. In the rural case this is straightforward as the assignment is not in general affected. In the urban scene the effect of a new project may cause major redistribution of the traffic-flow pattern.

The road-user benefits may then be written

$$B = C_O - C_N.$$

There are other complications, not generally apparent in the rural situation, viz.

(a) If a time stream of estimated future benefits is required then in the urban problem it is necessary to carry out an assignment year by year—an unhappy prospect!

(b) The future traffic that may be accommodated on the new system may not be able to be accommodated on the existing one. Theoretically this may be taken into account by resorting to the consumer-surplus notion of benefits and defining them in terms of the average equilibrium flows on the new and the old networks. If, however, the demand is relatively inelastic then $\frac{1}{2}(q_N+q_O)$ is near enough to $q_N$. It is still necessary, however, to take account of saturation of the old facility.

By way of final comment concerning this extremely complex matter of specifying benefits it should be pointed out that a good deal of current work is going on in the distribution of hourly volumes throughout the year.[99–101] At the present time a good approximation may be made by considering the urban-flow pattern to be classified into three volume classes—high, medium and low corresponding roughly to peak (including some high recreational), off-peak and certain week-end conditions and the 11 p.m.–6 a.m. period.

## ECONOMIC EVALUATION PROCEDURES

The manner in which the economic indices and parameters described above may be applied depends to a large extent on the purpose of the evaluation. The main purposes are

(a) to rank projects,
(b) to obtain the maximum benefits for a given budget,
(c) determination of desirable budget levels.

The methodology is also influenced by the nature of the problem particularly in relation to interdependence of the projects and the budget distribution in time.

Lorie and Savage[102] give a good appraisal of the standard procedures for dealing with (a), (b) and (c) above. Stimulated by their paper Weingartner undertook a major investigation into the application of linear programming to the capital budgeting problem area. A full account of his work is given in his book.[103] It is not intended to attempt here a review of the extensive literature and the well-developed methodology of the economic analyst. The purpose of this work, however, has been to produce a framework within which engineers, planners, economists, mathematical and other specialists can orient their approaches to the problems of land use and transport. It is appropriate therefore in this final chapter to preserve the dichotomy that has characterized methodology of the preceding chapters, i.e. the deductive and the objective, and introduce the student to the linear programming formulation of the evaluation problem.

It is perhaps worth mentioning that one of the virtues of the programming approach is that irrespective of whether or not a formal solution is taken out, the setting up of a complex problem with many variables in programming form is an important aspect of planning science—particularly as the dual variables are measures of the sensitivity of the constraints.

## LINEAR PROGRAMMING FORMULATION OF THE EVALUATION PROBLEM

If $b_1, b_2, \ldots, b_n$ represent the present value of the benefits derived from projects $x_1, x_2, \ldots, x_n$ over given time periods; and if $c_{11}$, $c_{21}, \ldots, c_{n1}$ represent the present values of the costs incurred in budget period 1; $c_{12}, c_{22}, \ldots, c_{n2}$ in period 2 and $c_{1m}, c_{2m}, \ldots, c_{mn}$ budget $m$. Then if $C_1, C_2, \ldots, C_m$ are the budgets for periods $1, 2, \ldots, m$ we may state the problem of maximizing the worth of the project programme in the following terms:

Maximize

$$\sum_{r=1}^{n} b_r x_r.$$

Subject to

(1) budget constraints

$$\sum_{r=1}^{n} c_{rs} x_r \leqslant C_s \qquad s = 1, 2, \ldots m,$$

(b) project constraints

$$x_r < 1 \qquad r = 1, 2, \ldots n$$

the $x$'s being non-negative.

In this problem the dual variables are of special significance. The primal has $n+m$ constraints and $n$ variables. The dual will thus have $n+m$ variables and $n$ constraints. Let the $n$ variables corresponding to

the project be designated by $y$ and those corresponding to the budgets by $z$. The dual may then be written:
Minimize

$$\sum_{s=1}^{m} z_s C_s + \sum_{r=1}^{n} y_r$$

Subject to

$$\sum_{s=1}^{m} z_s c_{rs} + y_s \geqslant b_r \qquad r = 1, 2, \ldots n$$

with the $z$'s and $y$'s $> 0$.

The above surfeit of symbols is more readily appreciated by constructing the primal/dual matrix.

TABLE 10.2. THE PRIMAL/DUAL MATRIX

| D \ P | $x_1$ | $x_2$ | ... | $x_n$ | |
|---|---|---|---|---|---|
| $z_1$ | $c_{11}$ | $c_{21}$ | | $c_{1n}$ | $\leqq C_1$ |
| $z_2$ | $c_{21}$ | $c_{22}$ | | $c_{2n}$ | $\leqq C_2$ |
| . | | | | | |
| . | | | | | |
| . | | | | | |
| . | | | | | |
| $z_m$ | $c_{m1}$ | | | $c_{mn}$ | $\leqq C_m$ |
| $y_1$ | 1 | | | | $\leqq 1$ |
| $y_2$ | | 1 | | | $\leqq 1$ |
| . | | | | | |
| . | | | | | |
| $y_n$ | | | | 1 | $\leqq 1$ |
| | $\geqq b_1$ | $\geqq b_2$ | | $\geqq b_n$ | |

In the optimal solution the $z$'s represent the marginal rates of return on the respective budgets and the optimal values of the $y$'s are the present worths of the individual projects.

The method therefore selects the worthwhile projects, provides a means of ranking them and indicates the "value" of the capital to be invested—a quite remarkable triple feat!

### *Independent Projects*

In the case of a set of independent projects with a single budget allocation the method is perhaps a little sophisticated and there are a number of much simpler methods that may be applied, viz.

(a) the projects may be ranked in order of their internal rates of return and the investment costs accumulated until the budget is matched;

(b) the projects may be discounted at given discount rates (as specified by a treasury or a board) and the benefit/cost ratios calculated. All ratios greater than unity are acceptable.

### *Mutually Exclusive Projects*

A discount rate (or minimum acceptable rate of return) needs to be specified. The project with the highest present worth (or benefit/cost ratio) is to be preferred.

### *Multiple Budgets*

With multiple budgets it is necessary to resort to more fundamental analysis. In a multiple-budget situation one may define the net present value or present worth of a set of independent projects as

$$w_1 = b_1 - y_1c_{11} - y_2c_{12} \cdots y_mc_{1m}$$

$$w_2 = b_2 - y_1c_{21} - y_2c_{22} \cdots y_mc_{2m}$$

$$\cdot \quad \cdot \quad \cdot \quad \cdot \quad \cdot \quad \cdot \quad \cdot \quad \cdot \quad \cdot \quad \cdot \quad \cdot \quad \cdot \quad \cdot$$

$$w_n = b_n - y_1c_{n1} - y_2c_{n2} \cdots y_mc_{nm}$$

It is clearly necessary to select a set of $y$'s that permit each of the budgets to be fully or nearly fully utilized. When this is done the $w$'s that have positive values are acceptable. This is, of course, the L.P. problem described above. However, a trial-and-error graphical approach suggested by Weingartner is instructive for a simple two-budget period case. Consider Fig. 10.1.

The lines

$$w_r = b_r - y_1 c_{r1} - y_2 c_{r2}$$

can be plotted as shown. The value of $w_s$ is proportional to the distance of these project lines from the origin. If a line through an arbitrarily chosen point $(y_1, y_2)$ is joined to origin then the projects may be ranked

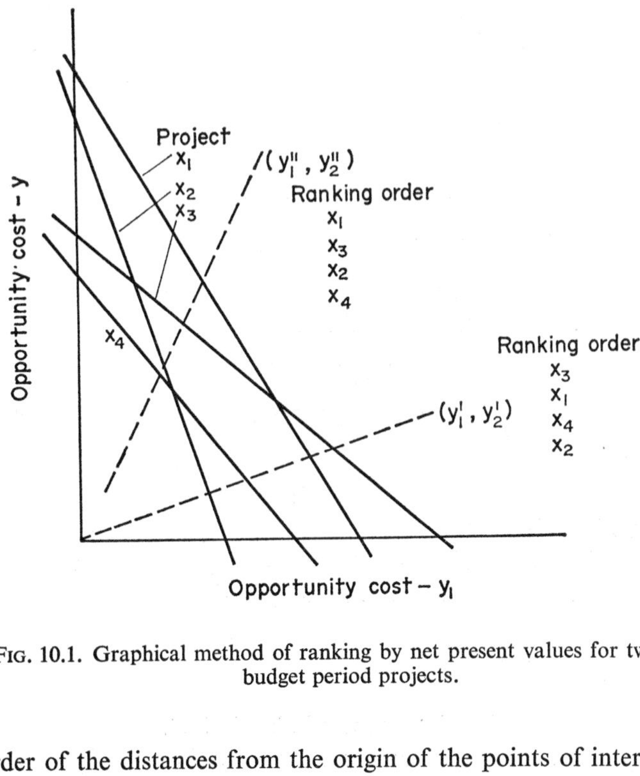


FIG. 10.1. Graphical method of ranking by net present values for two-budget period projects.

in order of the distances from the origin of the points of intersection of their project lines with the line from the origin to $(y_1, y_2)$. The two sets of investment costs are accumulated until one or other reaches the budget ceiling. If both approach it simultaneously then $y_1$ and $y_2$ are optimal. If not, new values of $y_1$ and $y_2$ are selected and the process repeated.

***Interdependent Projects***

This is perhaps the most important case for the engineer and planner. In complex situations one can seldom proceed directly to the optimum solution from basic physical and economic inputs. The best one can hope for is a set of alternative projects or project sets so that some best combination of them can be determined. In doing this it is most desirable that a number of alternative designs or solutions for certain projects be put forward. There is a need to consider contingent projects.

In this case trial and error methods may be used to construct a series of subsets that are comprised of independent projects. For example, if in a set of projects $A$, $B$, $C$, $D$, $E$ project $A$ may be undertaken in three ways $A_1$, $A_2$ and $A_3$, then we could investigate separately the three subsets $A_1, B, C, D, E$; $A_2, B, C, D, E$ and so on.

The linear programming formulation can take care of these additional constraints very simply, viz.

for mutually exclusive projects $x_1$, $x_2$ and $x_3$ we can ensure that only one is selected by introducing the constraint

$$x_1+x_2+x_3 \leqslant 1.$$

And, if $x_1$ is conditional on $x_2$ being undertaken then the further constraint

$$x_1 \leqslant x_2$$

must be introduced.

Most of the above points may be illustrated by the following simple example based on a problem discussed by McKean.[104]

## An Example

Consider the set of projects shown in Table 10.3.

Let us assume that $x_1$ and $x_2$ are mutually exclusive and further that the total budget available is $200,000.

McKean shows by heuristic argument that if the selection criterion is the maximization of present worth then projects $x_1$ and $x_3$ are the correct ones, not $x_2$ and $x_3$ as would be the case if the selection criterion were based on the internal rate of return.

It is left to the student to check the rate of return and the present value calculations, noting that the discounting rate is determined by marginal rate of return of 3%.

TABLE 10.3.

| Project | Investment $000's | Benefits Year I $000's | Benefits Year II $000's | Internal rate of return | *PVB* $000's |
|---|---|---|---|---|---|
| $x_1$ | 100 | 0 | 115 | 7 | 108 |
| $x_2$ | 100 | 110 | 0 | 10 | 107 |
| $x_3$ | 100 | 104 | 0 | 4 | 101 |
| Others returning 3% or less | | | | | |

The L.P. primal may be written down.

The budget constraint gives

$$100x_1+100x_2+100x_3 \leqslant 200.$$

For the mutually exclusive projects $x_1$ and $x_2$

$$x_1+x_2 \leqslant 1.$$

For independent projects

$$x_3 < 1.$$

The objective function

$$Z = 108x_1+107x_2+101x_3.$$

At this stage it is convenient to construct the primal/dual conversion matrix. It should be noted that there are only two independent projects (Table 10.4).

We can thus write the dual formulation

$$100y_1+z_1 \geqslant 108,$$
$$100y_1+z_2 \geqslant 101$$

and the objective function (to be minimized) is

$$Y = 200y_1 + z_1 + z_2.$$

The solutions of the primal and dual are

| Primal | Dual |
|---|---|
| $x_1 = 1$ | $y_1 = 1{\cdot}01$ |
| $x_2 = 0$ | $z_1 = 7$ |
| $x_3 = 1$ | $z_2 = 0$ |

TABLE 10.4.

| D \ P | $x_1$ | $x_2$ | $x_3$ | |
|---|---|---|---|---|
| $y_1$ | 100 | 100 | 100 | $\leqq 200$ |
| $z_1$ | 1 | 1 | | $\leqq 1$ |
| $z_2$ | | | 1 | $\leqq 1$ |
| | $\geqq 108$ | $\geqq 107$ | $\geqq 101$ | |

The results agree with McKean so far as the selection of projects is concerned. The opportunity cost of the capital is 1·01 and with this value the present worths of the selected projects are 7 and 0 respectively.

It should be stated here that it would be more correct to have set up the capital budgeting problem as an integer programme. The constraints for the $x$'s should be

$$x_r = 0 \quad \text{or} \quad 1.$$

The L.P. simplication does not, however, introduce undue complication unless the number of budget constraints is large. It can be shown that once a "good" project is under "treatment" its $x$ value will automatically rise to the upper bound, i.e. 1. However, due to the discrete nature of the investment costs any given set will not exactly sum to the total budget for the period. There is thus a residue in each budget period and the method seeks to use it up with fractional projects.

## IMPLEMENTATION

It seems appropriate to direct the final paragraphs of this work to some brief observations on implementation. If the evaluation model has taken realistic account of the budget constraints it would be fair to say that implementation should follow as a matter of course. Whilst this is generally true of the American approach—the 1956 Interstate Highway Programme provides a spectacular example—the fate of the Roskill Commission's recommendation on the siting of the third London airport provides salutary evidence to the contrary. But it is not intended to dwell here on the belated recognition of the overriding importance of certain environmental, social and political factors. Our concern throughout this book has been with the utility of quantitatively based studies and assessments to aid the decision-making process.

But whether or not implementation may be considered to be implicit in the final "economically" tested plan there now arise new problems relating to the process of implementation. They result from the grand scale of land-use and transport undertakings and their extended gestation periods. The implementation process is intimately concerned with time sequencing and dynamic programming. The methodological tools of operations research provide, as we have already seen, facility for this kind of study. Their formal exploitation is, however, being undertaken by the student of management and business science. Critical path scheduling, resource allocation, stock and inventory control, computer application, construction technology are the new specializations. Although detailed discussion of them is beyond the scope of this work, it is important to emphasize in these concluding words the closeness of the link between the planning and implementation phases. There are important interactions to recognise, e.g.

(a) efficient execution of the physical works will lower the investment costs and thus influence the economic evaluation:

(b) the order of carrying out separate parts of major land-use/transport schemes will alter the time pattern of the benefit and cost streams; this is particularly important in respect of operational consequences of transport components:

(c) the plans and strategies of the planning process should be the *more* rather than the *less* implementable:

(d) the basic methodology of systems planning and dynamic programming have a good deal in common.

There is, however, much yet to be done in training and practicing the new generation of systems analysts and planners before we can expect to achieve an overall optimization of the enormous range of factors and variables that govern and control the physical environment in which we live. In the meantime it is hoped that this work has produced some new vocabulary that not only will permit planners, engineers, and other specialists to converse more freely but will stimulate and guide the student to deeper reading of the expanding literature on systems analysis and planning.

## Reference Bibliography


1. CALIFORNIA DIVISION OF HIGHWAYS. *Los Angeles Regional Transportation Study*. A Prospectus prepared by Fairman, G. W., Advance Planning Department LARTS, September (1960).
2. BLUNDEN, W. R. *Introduction to Traffic Science*. London, Printerhall Ltd. (1967).
3. TANNER, J. C. *Factors Affecting the Amount of Travel*. Road Research Technical Paper No. 51, London, H.M.S.O. (1961).
4. BROCKMEYER, E., HALSTROM, H. L. and JENSEN, A. *The Life and Works of A. K. Erlang*. ACTA Polytechnica Scandinavica, Applied Mathematics and Computing Machinery Series No. 6, Copenhagen (1960).
5. MORSE, P. M. *Queues, Inventories and Maintenance*. John Wiley & Sons, Inc., New York (1958).
6. SAATY, T. L. *Elements of Queueing Theory with Applications*. McGraw Hill, New York (1961).
7. COX, D. R. The statistical analysis of congestion, *Journal Royal Statistical Society*, Series A, Vol. 118, pp. 324–35 (1955).
8. EDIE, L. C. Traffic delays at toll booths, *Operations Research*, Vol. 2, No. 2 (1954).
9. BUCKLEY, D. J. and BLUNDEN, W. R. Some delay flow characteristics for conflicting traffic streams, *Proceedings Second International Symposium on the Theory of Traffic Flow*, The Organization for Economic Cooperation and Development, Paris (1965).
10. KENDALL, D. G. Some problems in the theory of queues, *Journal Royal Statistical Society*, Series B, Vol. 13 (1951).
11. WINSTEN, C. B. Road and intersection capacity, in *Studies in the Economics of Transportation* by BECKMAN, MCQUIRE and WINSTEN, Yale University Press (1956).
12. PEARCY, T. Delays in landing air traffic, *Journal Royal Aeronautical Society*, Vol. **52**, p. 450 (1948).
13. BOWEN, E. G. and PEARCY, T. Delays in the flow of air traffic, *Journal Royal Aeronautical Society*, Vol. 52, p. 447 (1948).
14. LEE, A. M. and LONGTON, P. A. Queueing processes associated with airline passenger check-in, *Operational Research Quarterly*, Vol. 10, No. 1 (1959).
15. KOMETANI, E. An abridged table for infinite queues, *Operations Research*, Vol. 3, No. 4, p. 412 (1955).
16. LEE, A. M. *Applied Queueing Theory*. Macmillan, London (1966).
17. BUCKLEY, D. J. Road traffic headway distributions, *Proceedings Australian Road Research Board*, Vol. 1, Part 1 (1962).

18. RALLIS, T. *Capacity of Transport Centres—Ports, Railway Stations, Road Haulage Centres and Airports.* Report 35 Department of Road Construction, Transportation Engineering and Town Planning, Technical University of Denmark (1967).
19. BLUNDEN, W. R. On the traffic effects of frontage land uses on urban main roads and arterials, *Proceedings Australian Road Research Board,* Vol. 3, Part 1 (1966).
20. BLUNDEN, W. R., CLISSOLD, C. M. and FISHER, R. B. Distribution of acceptance gaps for crossing and turning manoeuvres, *Proceedings Australian Road Research Board,* Vol. 1, Part 1 (1962).
21. TANNER, J. C. The delay to pedestrians crossing a road, *Biometrika,* Vol. 38 (1951).
22. GAZIS, D. C., NEWELL, G. F., WARREN, P. and WEISS, G. H. The delay problem for crossing an *n*-lane highway, *Proceedings of Third International Symposium on Theory of Traffic Flow, Vehicular Traffic Science,* Elsevier Publishing Co. Inc., New York (1967).
23. JONES, P. G. *The Distribution of Acceptance Gaps for Crossing Multilane Traffic Flows.* M.Eng.Sci. Thesis, School of Traffic Engineering, University of New South Wales (1965).
24. ADAMS, W. F. Road traffic considered as a random time series, *Journal of Institution of Civil Engineers,* London, Vol. 4 (1936).
25. TANNER, J. C. *Theoretical Analysis of Delays at Uncontrolled Intersections.* R.R.L. Research Note RN3937/JCT, Road Research Laboratory, England (1961).
26. MAJOR, N. G. and BUCKLEY, D. J. Entry to a traffic stream, *Proceedings Australian Road Research Board,* Vol. 1, Part 1 (1962).
27. RICHARDS, P. I. Shock waves on the highway, *Operations Research,* Vol. 4, No. 1 (1956).
28. WEBSTER, F. V. *Traffic Signals Settings.* R.R.L. Technical Paper 39, London (1958).
29. MILLER, A. J. Settings for fixed-cycle traffic signals, *Proceedings Australian Road Research Board,* Vol. 2, Part 1 (1964).
30. NEWELL, G. F. Queues for fixed-cycle traffic lights, *Annals of Mathematical Statistics,* Vol. 31, No. 3, p. 589 (1960).
31. LEONG, H. J. W. Some aspects of urban intersection capacity, *Proc. Australian Road Research Board,* Vol. 2, Part 1 (1964).
32. MILLER, A. J. *The Capacity of Signalized Intersections in Australia.* Australian Road Research Board Bulletin No. 3 (1968).
33. BLUNDEN, W. R. and PRETTY, R. L. On deterministic cyclic flows in networks, *Second International Symposium on the Theory of Traffic Flow,* Organization for Economic Cooperation and Development, Paris (1963).
34. PRETTY, R. L. *Simulation of Traffic Flow in One- and Two-dimensional Road Layouts.* Ph.D. Thesis, School of Traffic Engineering, University of New South Wales (1966).
35. PRETTY, R. L. and BLUNDEN, W. R. On the computer simulation of a single channel queueing facility for a wide range of arrival and departure distributions, *Proceedings Australian Road Research Board,* Vol. 2, Part 1 (1964).
36. DAVIDSON, K. B. A flow travel-time relationship for use in transport planning, *Proceedings Australian Road Research Board,* Vol. 3, Part 1 (1966).

37. WARDROP, J. G. Some theoretical aspects of road traffic research, *Proceedings of Institution of Civil Engineers*, 1, London (1952).
38. DANTZIG, G. B. *Linear Programming and Extensions*. Princeton University Press, New Jersey (1963).
39. VAJDA, S. *Mathematical Programming*. Addison Wesley Publishing Co. Inc., London (1961).
40. HADLEY, G. *Linear Programming*. Addison Wesley Publishing Co. Inc., London (1967).
41. SIMONNARD, M. (translated by JEWELL, W. S.) *Linear Programming*. Prentice Hall Inc., New Jersey (1966).
42. FORD, L. R. and FULKERSON, D. R. *Flows in Networks*. Princeton University Press, New Jersey (1962).
43. DANTZIG, G. B. *Maximization of a Linear Function of Variables Subject to Linear Inequalities*. Chapter 21 in Monograph No. 13, Cowles Commission (editor KOOPMANS, T. C.), New York (1951).
44. DANTZIG, G. B. Application of the simplex method to a transportation problem, Chapter 23 in Monograph No. 13, Cowles Commission (editor KOOPMANS, T. C.), New York (1951).
45. FORD, L. R. and FULKERSON, D. R. Constructing maximum dynamic flows from static flows, *Operations Research*, Vol. 6 (1958).
46. WHITING, P. D. and HILLIER, J. A. A method of finding the shortest route through a road network, *Operations Research Quarterly*, Vol. 11 (1960).
47. BELLMAN, R. and KALABA, R. On *K*th best policies, *Journal Society of Industrial and Applied Mathematics*, Vol. 8 (1960).
48. HOFFMAN, W. and PAVLEY, R. A method for the solution of the *N*th best path problem, *Journal Association for Computing Machinery*, Vol. 6 (1959).
49. BLUNDEN, W. R., BUCHANAN, R. H. and OEDGOE, D. Costs of distribution, Chapter 13 in *Costs and Economics of the Australian Process Industries* (editors BUCHANAN, R. H. and SINCLAIR, C. G.), West Publishing Corporation Pty. Ltd., Sydney (1964).
50. R. J. SMEED. The Traffic Problem in Towns, *Traffic Eng. and Control*, **VI**, Nos. 6 & 7 (1964).
51. HUGHES, D. J. G. *A Study of the Effects of Land-use Density on Interzonal Travel Times by Public Transport*. M.Eng.Sci. Thesis, School of Traffic Engineering, University of New South Wales (1969).
52. SMITH, D. N. *Factors Affecting the Efficiency of Inter Modal Transport Operations*. M.Eng.Sci. Thesis, School of Traffic Engineering, University of New South Wales (1969).
53. PELENSKY, E., JAMES, H. and BLUNDEN, W. R. Economics of road improvements, *Proc. Australian Road Research Board*, Vol. 1, Part 1 (1962).
54. PELENSKY, E., BLUNDEN, W. R. and MUNRO, R. D. Operating costs of cars in urban areas, *Proc. Australian Road Research Board*, Vol. 4, Part 1 (1968).
55. JONES, J. H. and BLUNDEN, W. R. Ship turn-around time at the Port of Bangkok, *Journal of Waterways and Harbour Division, Proceedings of American Society of Civil Engineers*, Vol. 94, No. WW2 (1968).
56. HELLY, W. Efficiency in road traffic flow, *Proceedings of Second International Symposium on the Theory of Traffic Flow*, Organization for Economic Co-operation and Development, Paris (1965).

57. HILL, R. O'N. *On the Application of Queueing Theory to Transport Terminal Design.* M.Eng.Sci. Thesis, School of Traffic Engineering, University of New South Wales (1969).
58. CLARK, C. *Population Growth of Land Use.* Macmillan, London (1968).
59. VOORHEES, A. M. *Development Patterns in American Cities.* Highway Research Board Bulletin No. 293 (1961).
60. SWERDLOFF, C. N. and STOWERS, J. R. Tests on some first generation residential land-use models, *Public Roads,* Vol. 34, No. 5 (1964).
61. COLSTON, M. and BLUNDEN, W. R. On the duality of desire-line and land-use models, *Proceedings Australian Road Research Board,* Vol. 5, Part 1 (in preparation).
62. MUNT, P. W. and WEBSTER, F. V. *A Study of Road Traffic Crossing the Cordons Round Central Areas of Eight Towns with Populations Ranging from* 18000 *to* 8 *Million.* Road Research Laboratory, Ministry of Transport, R.R.L. Report LR209 (1968).
63. MERTZ, W. L. and HAMNER, L. B. A study of factors relating to urban travel, *Public Roads,* Vol. 9, No. 7 (1957).
64. SCHMIDT, R. E. and CAMPBELL, M. E. *Highway Traffic Estimation.* Eno Foundation, Saugatuck, Connecticut (1956).
65. BEESLEY, M. E. and KAIN J. F. Forecasting car ownership and use, *Urban Studies, 2* (1965).
66. KNEEBONE, D. C. and MUNRO, R. D. Work trips generated by a large city office building, *Proceedings Australian Road Research Board,* Vol. 2, Part 1 (1964).
67. MUNRO, R. D. Getting to work in Sydney—a point of concentration survey, *Proceedings Australian Road Research Board,* Vol. 4, Part 1 (1968).
68. FAIRTHORNE, D. B. Description and shortcomings of some urban road traffic models, *Operational Research Quarterly,* Vol. 15, No. 1 (1964).
69. KOMETANI, E. Estimation of origin destination trips using a transition matrix method, *Proceedings Third International Symposium on Theory of Traffic Flow, Vehicular Traffic Science* (editor EDIE, L. C.), Elsevier, New York (1967).
70. HANSEN, W. G. *Evaluation of Gravity Model Trip Distribution Procedures.* Highway Research Board Bulletin No. 347 (1962).
71. WYNN, F. H. and LINDER, C. E. *Tests on Interactance Formulae derived from O–D Data.* Highway Research Board Bulletin No. 253 (1960).
72. SCHWARZ, H. Methods for determining trip distribution, *Traffic Engineering and Control,* Vol. 4, No. 11 (1963).
73. MUNRO, R. D. Journeys to work, *The Chartered Surveyor,* Vol. 97, 10 (1965).
74. COHEN, M. H. The relative distribution of households and places of work—A discussion on a Paper by Wardrop, J. C., *Proceedings of a Symposium on Theory of Traffic Flow,* Elsevier, New York (1961).
75. STOUFFER, S. A. Intervening opportunities—a theory relating mobility and distance, *American Sociological Review,* December (1940).
76. SCHNIEDER, M. *Chicago Area Transportation Study.* Final Report, Vol. 2 (1960).
77. TOMAZINIS, A. R. *A New Method of Trip Distribution in an Urban Area.* Highway Research Board Bulletin No. 347 (1962).
78. COMMONWEALTH STATISTICIAN. *The Journey to Work.* State Planning Authority of New South Wales (1965).

79. BELL, G. *Factors Affecting the Distribution of Work Trips in Urban Areas.* M.Eng.Sci. Thesis, School of Traffic Engineering, University of New South Wales (1966).
80. CONNORS, G. H. *Prediction of Traffic Desire Lines by Linear Programming Using Real Data.* M.Eng.Sci. Thesis, School of Traffic Engineering, University of New South Wales (1968).
81. MCDONALD, W. R. and BLUNDEN, W. R. The application of linear programming to the determination of road traffic desire-line patterns, *Proceedings of Australian Road Research Board,* Vol. 4, Part 1 (1968).
82. CARROL, J. D. The relation of homes to work and the spatial pattern of cities, *Social Forces,* Vol. 30, No. 3 (1952).
83. BELL, G. The distribution of work trips in urban areas—some considerations, *Proceedings Australian Road Research Board,* Vol. 5, Part 1 (in preparation).
84. IRWIN, N. A. and VON CUBE, H. G. *Capacity Restraint in Multi-travel Mode Assignment Programs.* Highway Research Board Bulletin No. 347 (1962).
85. TOMLIN, J. A. A linear programming model for the assignment of traffic. *Proceedings Australian Road Research Board,* Vol. 3, Part 1 (1966).
86. MEYER, J. R., KAIN, J. F. and WOHL, M. *The Urban Transportation Problem.* Harvard University Press, Cambridge, Mass. (1965).
87. ALMOND, J. Traffic assignment with flow-dependent journey times, *Proceedings of Third International Symposium on Theory of Traffic Flow, Vehicular Traffic Science* (editor EDIE, L. C.), Elsevier, New York (1967).
88. WARDROP, J. G. The distribution of traffic on a road system, *Proceedings of First International Symposium on the Theory of Traffic Flow,* Elsevier, New York (1961).
89. KIRBY, R. F. and POTTS, R. B. The minimum route problem for networks with turn penalties and prohibitions, *Transportation Research,* Vol. 3, No. 3 (1969).
90. JEWELL, W. S. Models for traffic assignment, *Transportation Research,* Vol. 1, No. 1 (1967).
91. HUBER, M. J., SOUTHWELL, H. B. and WHITEFORD, D. K. *Comparative Analysis of Traffic Assignment Techniques with Actual Highway Use.* National Cooperative Highway Research Program Report 58, Highway Research Board (1968).
92. FARRINGTON, P. G. *On the Application of Current Theories of Traffic Assignment to "Real" Road Systems.* M.Eng.Sci. Thesis, School of Traffic Engineering, University of New South Wales (1967).
93. SUB-COMMITTEE ON BENEFITS AND COSTS. *Proposed Practices for Economic Analysis of River Basin Projects.* Report to the Federal Inter-Agency River Basin Committee, Washington (1950).
94. COMMITTEE ON PLANNING AND DESIGN POLICIES. *Road User Benefit Analysis for Highway Improvements.* A.A.S.H.O., Washington (1960).
95. DAWSON, R. F. F. *The Economic Assessment of Road Improvement Schemes.* Road Research Laboratory, Ministry of Transport, Road Research Technical Paper No. 75, H.M.S.O., London (1968).
96. BEESLEY, M. E. and REYNOLDS, D. J. *The London–Birmingham Motorway: Part II. Economic Assessment.* Road Research Laboratory, Ministry of Transport, Road Research Technical Paper No. 46, H.M.S.O., London (1960).
97. LACK, G. N. T., DELANEY, D. J., FISHER, N. W. F., SPENCE, J. A. and THOMPSON, K. E. A model for the economic evaluation of rural road improvements, *Proceedings Australian Road Research Board,* Vol. 4, Part 1 (1968).

98. THOMPSON, K. E., DELANEY, D. J. and LEES, P. A model for the economic evaluation of urban road projects, *Proceedings Australian Road Research Board*, Vol. 5, Part 1 (in preparation).
99. VAUGHAN, R. J. *The Distribution of Hourly Traffic Volumes and its Application to Economic Design Concepts*. M.E. Thesis, School of Traffic Engineering, University of New South Wales (1969).
100. VAUGHAN, R. J. The estimation of highest hourly volumes, *Proceedings Australian Road Research Board*, Vol. 4, Part 1 (1968).
101. SOUTHWELL, P. G. *A Study of Factors Influencing the Hourly Distribution of Traffic Flows*. M.Eng.Sci. Thesis, School of Traffic Engineering, University of New South Wales (in preparation).
102. LORIE, J. H. and SAVAGE, L. J. Three problems in rationing capital, *The Journal of Business*, University of Chicago, School of Business, Vol. 28, No. 4 (1955).
103. WEINGARTNER, H. M. *Mathematical Programming and the Analysis of Capital Budgeting Problems*. Prentice-Hall (1963).
104. MCKEAN, R. N. *Efficiency in Government through Systems Analysis*. John Wiley & Sons Inc., New York (1958).

# INDEX